ELECTRIC MOTORS IN THE
HOME WORKSHOP

Electric Motors in the Home Workshop

A practical guide to methods of utilising readily available
electric motors in typical small workshop applications

Jim Cox

SPECIAL INTEREST MODEL BOOKS
Special Interest Model Books Ltd.
P.O. Box 327
Poole
Dorset
BH15 2RG

First published by Nexus Special Interests Ltd. 1996

Reprinted 1999

© Jim Cox 1996

This edition © 2002 Special Interest Model Books Ltd.

ISBN 1-85486-133-6

Phototypesetting by The Studio, Exeter
Printed and bound in Great Britain by Biddles Ltd, *www.biddles.co.uk*

Contents

Introduction

Electric Motors, No 16 in the Nexus Special Interests Workshop Practice Series, was written to provide basic information on the very wide range of motors and related devices that could be used in the home workshop environment. It has been widely read and is now in its fifth printing.

As a matter of general interest it also included some advice on how to utilise motors from discarded equipment and some simple methods of operating industrial three-phase motors from domestic single-phase supplies. It is clear from subsequent published and unpublished correspondence that these latter two subjects are of great interest to the model engineering community and there is a need for more comprehensive and detailed information.

This new book is intended to fill this gap. It includes information on the how's and why's of motor operation but the main part of the text is devoted to practical methods of identifying and obtaining satisfactory operation from discarded and surplus motors. Advice is given on suitable methods of speed control and, because many commutator motors are operated from battery supplies, a section is also included which covers the characteristics and use of primary and rechargeable cells.

FOREWORD

Safety

Electric power is in every home and is normally so safe and convenient that we tend to take it for granted. Providing the high voltages are protected by properly insulated cables and terminations, the chances of electric shocks are remote and this is the normal situation in home use. However, if live conductors are exposed or if the connected equipment is faulty, the normal 240V domestic mains supply is quite capable of giving the user a very unpleasant, or in extreme cases, fatal electric shock. It is essential therefore to take proper care and to observe safe working practices when wiring up, installing or testing mains voltage electric motors and associated equipment.

The following guidelines should always be observed:

1. Switch off **AND** unplug from the mains before touching any conductors that might be live in normal operation. If it is in a permanently wired circuit which cannot be unplugged then switch off and remove the fuse supplying that circuit. Keep the fuse in your pocket to ensure that no-one else can put it back before you are ready. It is **NOT** sufficient to just switch off without also disconnecting the circuit. As any insurance company will tell you, faulty switches are not particularly rare occurrences and you don't want to find out the hard way! In the case of permanently wired circuits it is also good practice to use the blade of an insulated screwdriver to short together the input conductors before you allow your fingers near them – just to make sure that you have switched off the right circuit and pulled the right fuse.

2. The strength of an electric shock is determined by the amount of current that flows and not directly by the voltage. In the human body most of the resistance to current flow is in the thin outer layer of dry skin. This covers our fat and muscle which are soaked in body fluids and because of this offer much less resistance to current flow. Anything which penetrates or wets (including perspiration) the skin greatly increases our susceptibility to electric shock, so always keep your hands dry and cover any cuts or abrasions.

A key point to bear in mind is to avoid the possibility of any electric shock where the current path is through the chest as this can upset the heart muscle. the common danger paths are, hand to hand, hand to foot and hand to head.

Damp concrete floors are quite good conductors, so if you are working on this sort of surface wear rubber soled shoes to break the hand to foot path.

Hand to head paths are relatively uncommon – most of us instinctively avoid touching things with our head. The main hazard here is the head touching the metal parts of an earthed bench light or reading lamp. The answer here is to use either a lamp with no exposed metal parts or a modern double insulated lamp. These have two-core instead of three core flex leads and the doubly insulated metal parts are left floating with no connection to the supply or ground.

Occasionally, professional electricians have to work on live circuits which cannot be disconnected. They are used to the meticulous care that this demands and one technique which is used is never to touch the work with more than one hand at a time – the other hand being kept safe not touching any metallic object. You are advised emphatically **NOT** to work on live circuits; however, where practicable, the principle of avoiding two-hand contact is a useful additional safeguard.

3. Always connect to earth/ground the motor frame or equipment casing – even on temporary test rigs. This is good practice with any equipment but doubly important on items of dubious origin where the fault may be an intermittent failure of the insulation between windings and frame!

4. Before applying power to a motor or similar device do make sure that it's properly anchored to something solid. When a motor starts, the reaction to the starting torque can cause it to leap off the bench with obvious electrical and mechanical hazards.

5. Be sure you understand the correct connections of the motor/circuit that you are working on. In subsequent chapters of this book advice is given on testing, operation and installation of motors in a clear and understandable form. However, it is not possible to anticipate all eventualities, so if you are in any doubt don't take risks – consult a qualified electrician.

6. At some stage in the initial installation and testing, live terminations may have to be accessible for voltage or current monitoring. Stringent observation of the precautions outlined in 1, 2 and 3 is necessary. Once this initial phase is complete, all possibly live points must be properly enclosed in an **EARTHED** metal casing and protected from the ingress of cutting fluids or contact with flammable materials, in extreme cases of failure it is possible for overheated motors or control equipment to present a fire risk. The metal casing of most motors normally contains this risk within the motor – the same should be true of any associated equipment.

7. The above comments refer particularly to equipment operating from 220/240 volt domestic mains supply. 115 volt supplies are more forgiving, as the hazardous currents are roughly halved, but still need to be treated with respect. Below 50 volts, shocks are rarely hazardous unless exceptionally low resistance contact is made or the individual is particularly susceptible to shocks of any kind.

Remember that if you are unlucky enough to experience a really bad shock the muscles of your hand and arm may contact so strongly that you cannot let go. Don't panic and try to fight it but step backwards so that your hands are dragged free of the danger area.

At 6 to 24 volts, which is typical of automobile and model activity, the main hazard is thermal as the power source may be capable of delivering large short circuit current which can raise connecting wires to red heat in an embarrassingly short time. If working on automobile battery or related circuits do not wear a wrist watch with a metal wristband. It is natural to rest the wrists on the equipment when making connections and if a metal wristband should happen to bridge terminations a very nasty burn could result in seconds.

8. Industrial motors are rated for maximum winding temperatures in the range 100°C/210°F to 165°C/330°F. The motor casing will be cooler than the windings and these temperatures are only reached at full load and maximum ambient temperature. Nevertheless, be careful, as there are plenty of occasions when the outside parts of a motor will be hot enough to deliver a nasty burn if grasped incautiously.

CHAPTER 1

Single-Phase Induction Motors

1.1 General

Before you start looking around for potentially useful motors it's useful to know a bit about the sort of motor you are looking for. Induction motors are the kind of motor you usually find fitted to the larger pieces of machinery in the home workshop, typically lathes, milling machines and pillar drills. Figure 1.1 shows some examples.

Single phase means that they are suitable for operation from normal domestic two wire plus earth, three pin 240V AC power points as distinct from three-phase induction motors which operate from three wire 415V industrial supplies. An AC supply is essential because induction motors will not operate from DC supplies or batteries.

The key characteristic which makes

Fig. 1.1 *Single-phase induction motors*

them so suitable for machine tool use is their comparatively low and constant shaft speed. One or two stages of belt or gear reduction is sufficient for most machines and wide variations in cutting load have little effect on cutting speed.

1.2 Construction

A disassembled motor is shown in Figure 1.2. The two vital parts are the rotor and the stator. The active part of both rotor and stator is a stack of electrical grade silicon steel laminations. The main windings which generate the rotating magnetic field are fitted into slots in the stator. The rotor has no separate winding but the slots in it are filled with copper or aluminium bars which are all connected together at the ends of the rotor by solid metal shorting rings.

In operation, the magnetic field from the stator induces large currents (this is why it's called an induction motor) in the rotor bars and the interaction between these currents and the stator field causes the rotor to rotate at a speed a little slower than the rotation of the magnetic field.

This speed of rotation is determined by the supply frequency (50Hz = 50 cycles per second) and the arrangement of the windings in the stator. With the main windings grouped into two sectors (usually called 'poles') the magnetic field rotates at 50 cycles per second = 3000 revolutions per minute (rpm) and the rotor speed will be about 95% of this i.e. 2850 rpm. Two-, four-and six-pole winding arrangements are common with corresponding rotor speeds of 2850 rpm, 1425 rpm and 950 rpm.

Fig. 1.2 *Disassembled motor*

Motors can be wound with larger numbers of poles to achieve lower operating speeds but both output power and efficiency suffer. They can also be wound with two or three sets of windings, each giving a different operating speed. The two pole/four pole combination (2850/1425 rpm) is popular because it gives a useful range of speeds and is not much larger than the equivalent single speed machine.

In this book the descriptions of the theoretical aspects of motor operation are kept to the minimum necessary to give a reasonable understanding of the recommended methods of utilising particular motor types. Fuller explanations can be found in my book *Electric Motors* (Workshop Practice Series No. 16, published by Nexus Special Interests).

1.3 Starting

It is unfortunate that an induction motor stator operating from the normal domestic single-phase supply produces a field that pulsates rather than rotates at supply frequency. Because of this, it cannot generate significant torque unless the rotor is first run up to about two-thirds normal operating speed. To achieve this, additional windings (start windings) are fitted to the stator, mainly occupying the empty or partly filled slots between the sectors of the main winding. By controlling the current flow in this winding, the field generated by the main winding is modified sufficiently to enable it to generate torque right down to zero speed. While this solves the starting problem, in most cases it is a very inefficient way of generating torque. Large currents are taken from the supply while the start winding is in circuit, and it has to be disconnected automatically by a centrifugal switch as soon as the motor reaches operating speed to prevent serious overheating.

The simplest and lowest cost method is used in 'split-phase' motors. In this case, the start winding consists of slightly fewer turns of thin wire that are connected in parallel with the main winding while the motor runs up to speed (see Fig. 1.3). This arrangement typically draws six times full load current from the supply while the motor runs up to speed, and can deliver a starting torque of about one to two times full load torque. It is unsuitable for frequent start/stop operation because this may cause the small start winding to overheat.

A better but more expensive system is used in 'capacitor-start' motors (see Fig. 1.4). In this case a capacitor is connected in series with the start winding. This improves the efficiency of the starting system and both increases the starting torque and reduces the starting current.

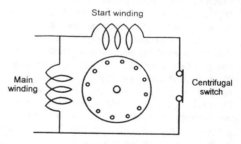

Fig. 1.3 *Split-phase motor*

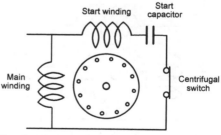

Fig. 1.4 *Capacitor-start motor*

In each case, a centrifugally operated switch (shown on the right-hand side of the two figures) keeps the start winding in circuit until the rotor is turning fast enough to be able to deliver at least full load torque. The centrifugal force at this speed is then high enough to open the switch contacts which then remain open as the rotor speed continues to increase to its normal operating speed.

Somewhat less common, but often found in fans, washing machines, dishwashers and 'hobby' grade workshop machinery, is the 'capacitor-run' type motor. This is not really a single-phase machine at all, but a two-phase machine where the second phase is provided by a capacitor that is in circuit all the time (see Fig. 1.5). The two sets of windings occupy approximately equal space in the stator slots. The capacitor windings frequently use more turns of finer wire as this permits the use of a smaller and less expensive capacitor. Because the capacitor-fed second phase provides torque right down to zero speed, the capacitor-run motor does not need a centrifugal starting switch, but the penalty is poor starting torque — sometimes less than half full load torque. Nevertheless, many items of workshop equipment can start up at either no load or light load and this sort of motor can be very suitable.

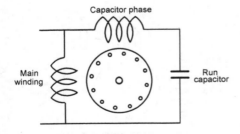

Fig. 1.5 *Capacitor-run motor*

1.4 Speed and torque
With the motor running light without any applied load, all three types almost reach synchronous speed (i.e. when the rotor rotates at the same speed as the magnetic field). As the load is increased, the speed drops by about 5% when the motor is delivering full rated horsepower for the split-phase and capacitor-start types, and by 8% to 10% for most capacitor-run types. As the load is increased further the speed drops, until at about two-thirds synchronous speed it reaches the maximum torque it can produce and is delivering typically twice its rated power.

Any increase in load beyond this point, even a very small momentary increase, will result in the motor torque being insufficient to turn the load so it will stall abruptly and come to a dead stop. This torque is known as the 'pull-out torque'. If the load on the motor exceeds this value the rotor stops turning and the centrifugal switch brings the start winding back into circuit. This attempts to restart the motor but, unless the load on the motor is immediately reduced (within seconds!), the motor will be unable to start and serious overheating will occur.

1.5 Power rating
Most industrial induction motors are rated to deliver their nameplate power continuously at local air temperatures of up to 40°C. In home workshop use temperatures are usually much lower, and periods of operation at maximum load are usually in minutes rather than hours. The motor never approaches its continuous full load maximum temperature rating.

Because of this, for reasonable periods of time, it is possible to operate in the overload region between the torque at rated output and peak torque where the motor may be delivering up to twice its rated power. The motor must never be allowed to stall because all the input power is then dissipated in the windings and overheating is very rapid indeed.

Many motors operate for most of the time at a small fraction of their rated power and their outside casing may be no more than comfortably warm. There is a temptation to assume that this is the normal operating temperature for the motor and there is something wrong if it runs hotter. This is far from true, especially with modern motors. Improvements in design have enabled more power to be delivered from a given motor size. Because there has been little change in efficiency, the same total losses inevitably heat the smaller motor to a higher temperature. The insulation materials used are designed to withstand these higher temperatures and, depending on motor design, maximum winding temperatures as high as 165°C/330°F may be normal. The temperature of the outside surface of the motor is typically 10°C to 20°C lower than this but still needs to be treated with respect. This is not just hot enough to fry eggs — it is hot enough for even casual contact to result in a nasty burn!

This ability to operate in the overload region for reasonable periods of time is very useful in the home workshop as it often makes it possible to use a smaller motor. Common major items are a 3½" centre height lathe and a small vertical mill. The manufacturer's recommendation for this size of machine is typically ¾ hp (560W), but this rating is based on the assumption that the machine may be used continuously with carbide-tipped tooling near the limit of its capacity. This can certainly happen in industrial use but in the amateur's workshop it is a fairly rare occurrence. In fact, if HSS tooling is used, it is difficult to utilise even half this

power without incurring unacceptably short tool life. My first machines in this class were an ML7 lathe and a Centec 2A mill — both were fitted with ¼ hp motors. These machines led a hard life in my workshop for more than ten years before I upgraded and the only occasions that I really missed the extra power was in heavy turning operations on large diameter stainless steel.

1.6 Picking a suitable motor

If brand new motors are bought, the above comments hardly apply because the cost saving in buying the smaller motor is not large and it makes sense to have the additional power for a relatively small increase in outlay. However, if the surplus market or the scrapyard is the intended source it can make a big difference. Old washing machines, mainframe computers and small industrial machinery are plentiful sources of single-phase induction motors in the power range ¼ hp to ½ hp and can be purchased for, at most, a few pounds each. Larger single-phase motors are comparatively rare because, if more than ½ hp is needed, the industrial user will almost always use a three-phase machine and domestic equipment turns to high speed commutator motors.

Obsolete or faulty industrial machinery is probably the best source of suitable motors. Large cooling fans, pumps, motor operated valves, bacon slicers, hoists and special purpose process control machines are often powered by fractional hp single-phase motors. These are usually foot mounted with belt drive so that they are easy to remove and re-install. Best of all is the fact that the motors are very rarely faulty. To the industrial user a motor is a fairly low cost item and easy to replace. In the comparatively rare cases where the motor fails it will normally be replaced rather than discard the whole machine. When the machine is finally scrapped it is usually for other reasons and it will still include a perfectly good motor with many years of useful life remaining.

The other side of the coin is to view any isolated second-hand motor on the scrap pile with considerable suspicion. If it's a good motor why isn't it still on its parent machine? Particularly unpromising is an apparently undamaged motor that has been carefully disconnected and removed from its original location. The only time that this normally occurs is when a maintenance engineer replaces a dud motor! Unless you can discover why it has been removed, leave well alone.

Paradoxically the motor that has been removed with the aid of bolt cutters and a cutting torch is a much better bet. These are the tools of the wrecker who is removing the motor to salvage its copper content. If the wires to the motor have been sheared off or cut, rather than disconnected, there is an excellent chance that the motor is good.

The good thing about industrial motors is that they usually carry an informative nameplate so, you've a reasonably clear idea of what you're getting.

The following information is usually provided:

Rating
CONT = continuous or INT = intermittent. If neither rating is shown the motor will normally be continuous rated.

Supply voltage
V = volts. More than one supply voltage may be shown if the motor is a dual voltage type. If the motor is a six-terminal three-phase type (see Chapter 2) supply voltages will be shown for both the star and delta connections.

Supply current
FLA = full load amps.

Supply frequency
Hz = Hertz = cycles per second.

Shaft speed
r/min or l/min or rpm = revs per minute.

Power factor
PF or Cos ϕ — this is the fraction of the load current which represents the true power consumed by the motor. The remainder is 'wattless' current caused by the motor windings accepting power over part of one cycle of the supply frequency and returning this power to the supply during the rest of that cycle. Domestic electricity meters record only the true power component and ignore the 'wattless' part.

This rating is only of interest to the large scale industrial user who wishes to control the overall power factor of his factory load.

Mechanical power
kW = kilowatts or hp = horsepower
1 hp = 0.746 kW, 1 kW = 1000 W.
Older motors are rated in hp. Modern motors, while still made in the original fractional and integral hp sizes, now express these ratings in kW, e.g.

¼ hp	0.18 kW	1 hp	0.75 kW
⅓ hp	0.25 kW	1½ hp	1.1 kW
½ hp	0.37 kW	2 hp	1.5 kW
¾ hp	0.56 kW	3 hp	2.2 kW

If the motor is a capacitor-start type, the capacitor will usually be bolted directly to the motor casing at the 10 o'clock or 2 o'clock position (Fig. 1.6). If it is a capacitor-run type, the capacitor may be mounted in this position but it is often mounted separately from the motor. Capacitors come in many different sizes and shapes — examples are shown in Chapter 8. It is important to locate any associated capacitors and take note of the connections — better still is to remove the motor with any capacitors and/or starting gear still connected. Also take a note of any nameplate on the parent machine — this information may be useful if spares are needed.

1.7 Sorting out the connections
Once the motor is safely home there may be a problem in sorting out the right connections. Unfortunately there is little standardisation on the arrangement of motor terminations and you may be faced with almost any combination of terminals, wires, colour codes and numbering systems. If you have managed to preserve enough of the original wiring this is easy but if you are faced with an anonymous set of terminals or pigtails a little detective work is needed.

Before applying any power to the motor make sure that the windings are not wet or damp and check out the insulation to frame as described in Chapter 4. With the insulation checked out now make a direct and reliable connection between the motor metalwork and the electricity supply earth. This is essential to ensure that, when power is eventually applied, insulation failure or a misconnection cannot result in 'live' metalwork.

Finally anchor it safely to the bench so that a sudden start or stop doesn't send it cavorting round the workshop.

1.8 Two terminations
The simplest, and fortunately the most common arrangement, is just two terminals with two wires connected to each terminal corresponding to the wiring arrangement shown in Figures 1.3 and 1.4. This is straightforward and if the motor starts and runs happily in the required direction when power is applied you are home and dry. However, pure

Fig. 1.6 *Capacitor-start motor*

cussedness will usually ensure that you need the opposite direction of rotation. Unlike three-phase machines, single-phase motors cannot be reversed while they are running and are sometimes thought to be unsuitable for operation with either direction of rotation. Fortunately this is not true; the construction of a single-phase motor is completely symmetrical and it will run equally well in either direction. The direction of rotation is determined by the relative direction of the current flow in the start and main windings during the initial starting period while the motor is running up to speed. If both currents flow in one direction, clockwise rotation will result. If the relation between the current flows is changed by reversing the connections to either (not both) the start or the main winding, the next time the motor is started, it will run in the reverse direction.

If the motor is to be used to power a lathe a reversing switch is a useful accessory, but it is only effective when the motor is stationary. If it is operated while the motor is running it will have no effect and the shaft will continue to rotate in the same direction. However, if power is removed and the motor allowed to stop, the next time power is applied it will start and run in the reverse direction.

Our unknown motor will probably not have any useful identification on the leads but in this simple case it is easy to sort them out by trial and error. Identify the leads as 1 and 2 connected to terminal A, 3 and 4 to terminal B. Reverse 1 and 3 so that the new grouping is 3 + 2 and 1 + 4. If you are lucky the motor will now start and run in the opposite direction. If not, no harm is done because the ends of each winding are connected to itself and no current can flow. Rearrange the connections to 3 + 1 and 2 + 4 and the motor will now run happily in the required direction.

1.9 Three terminations
Life gets more interesting as the number of connections increase. If there are three terminations, first check to see if a centrifugal switch is fitted. This is normally located on the non-drive end of the rotor shaft. If there is no switch, the motor is either:

> three phase or
> capacitor run or
> split phase with external relay.

A three-phase motor is easy to identify because it is completely symmetrical (Fig. 1.7) and the A − B, B − C and C − A winding resistances will all be equal. Three-phase motors are dealt with in Chapter 2.

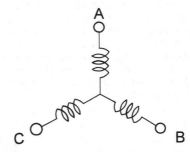

Fig. 1.7 *Three-phase winding*

The capacitor-run motor will usually have two approximately equal windings with one end commoned (Fig. 1.8). In this case the A − B and B − C resistances will be roughly equal and the C − A resistance about twice as much. Whichever is the lower of the A − B and B − C pairs is the main winding. The higher resistance pair is the capacitor-fed phase. The two sets of coils will be mounted 90 degrees apart on the stator and will be roughly similar in size i.e although one set may be wound with thinner wire it will have more turns so that the total volume of copper is about the same.

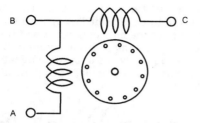

Fig. 1.8 *Capacitor-run winding*

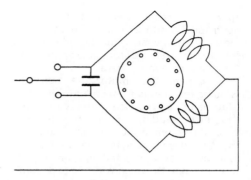

Fig. 1.9 *Capacitor-run reversing*

If you have the correct capacitor, connect as in Figure 1.5 and apply power. The motor should start and run smoothly and quietly. The torque pulsations in a capacitor-run motor are less than in a split-phase or capacitor-start machine and it is noticeably quieter. If the correct size of capacitor is unknown, start with about 8 μF and try to select a value which makes the voltage across the capacitor phase about equal to the supply voltage (because the capacitor and the inductance of the capacitor-fed phase form a resonant circuit, it is quite possible for the voltage across this phase to **exceed** the supply voltage). Dependent on motor size, the useful range is about 2 to 30 μF. Don't worry if you haven't got a large enough capacitor, a smaller value will do no harm. It will reduce the starting torque but the full-load torque will be only slightly affected.

The motor can be reversed by interchanging the connections to **either** the main or the capacitor-fed winding. If it is a symmetrical motor i.e. both windings of equal resistance, either winding can be used as the capacitor-fed phase and the very simple reversing arrangement shown in Figure 1.9 can be used.

If it is a split-phase motor the resistance of the start winding will be higher than that of the main winding. The start winding coils occupy stator slots 90 degrees away from the main winding coils. The start winding has about the same number of turns as the main winding but is wound with thinner wire so the total volume of copper is much less and the start winding slots are less than half full.

Connect as in Figure 1.10 with the lower resistance winding connected directly to the supply and the higher resistance winding connected via a starting push button. Press the start button, apply power and release the start button after one or two seconds. The motor should now run normally. It can be reversed in exactly the same way as a

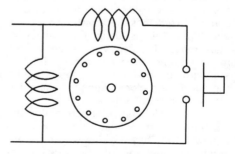

Fig. 1.10 *Split-phase starting*

13

split-phase motor fitted with an internal centrifugal starting switch. This type of motor is often used in the sealed units of freezers and refrigerators but, apart from these uses, is rather rare.

In a capacitor-run motor, the capacitor winding may range from an equal number of turns to nearly twice as many turns as the main winding. Somewhat thinner wire is used when the number of turns in the capacitor winding is large, so that the resistance of this winding may range from equal to the main winding to several times larger.

In extreme cases it is possible to confuse the two types of motor. If there is any doubt remove the end bell of the motor and examine the windings. With a split-phase motor the main winding will occupy most, if not all, of the stator slots. The start winding will be clearly different because it is wound with much thinner wire and inserted into empty or partly filled slots occupying a much smaller fraction of the total slot space.

With a capacitor-run motor there will be little or no difference in the thickness of the wire used for the two windings. There will be no marked difference between the two windings and each will occupy about the same total amount of slot space.

The last variation on the three-termination motor is the type with a centrifugal switch fitted in series with one of the windings. This is usually a capacitor-start motor for use with a separately mounted capacitor or, more often, a standard capacitor-start motor where the normal 2 o'clock capacitor has fallen off or been removed.

Three possible ways of connecting the capacitor start winding are shown in Figure 1.11. They are all straightforward series connections of the three components and are exactly equivalent. It doesn't matter which one is used but bear the three possibilities in mind when sorting out the connections. A and B are the most common arrangements; C is rare because it needs a fourth termination and has no particular compensating advantage.

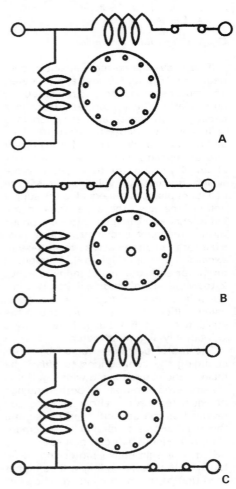

Fig. 1.11 *Capacitor-start wiring*

14

Capacitor-start motors need much higher values of capacitance than capacitor-run types. Capacitors *must* be suitable for AC working but, because they are in circuit for only a few seconds at a time, special intermittent rated AC electrolytic capacitors can be used (see Chapter 8). These pack a large value of capacitance into a small volume. The full-load torque and output power are independent of capacitance value which affects starting torque only. The useful range is about 15 μF to 150 μF. Low values are fine if you are not too fussy about starting torque. If you have to make a guess at good value for a motor in the ¼ hp to ½ hp range try 50 μF for the smaller motor and 100 μF for the larger.

If there is no prior information it can be difficult to distinguish between capacitor-start and split-phase motors because the winding arrangements are so similar. For the same power rating, the main windings are identical. In a capacitor-start motor the optimum number of turns and the resistance of the start winding depends on the capacitor value chosen for the design. Usually the winding has rather more turns than the equivalent split-phase machine, but if the design is based on a rather large value of start capacitor, the start winding can be very similar to the equivalent split-phase design.

The good news is that it may not matter. If the capacitor is omitted and the start winding connected directly to the centrifugal switch, some types of capacitor-start motor will start and run as a split-phase machine. The starting torque will be poor and the starting current high, but the full-load performance will be unaffected.

The converse is also true. If a suitable (i.e. about the same capacity as that needed for a similar power rating capacitor-start motor) capacitor is inserted in series with the start winding of a split phase motor, the starting torque will usually be improved and the starting current reduced. The full-load performance will be unchanged.

This interchangeability of function only applies to split-phase and some types of capacitor-start machines because of the similarity of their start windings. Capacitor-run machines cannot operate in this way and must always be fitted with a suitable run capacitor.

1.10 Four terminations
Four terminations are the next step in complexity. These are usually the same as the two-termination motors discussed earlier in section 1.8 but with each of the four wires brought out to a separate termination for convenient connection to a reversing switch. The two termination methods can be used to sort out the connections.

A different type of four-termination motor is a dual-voltage, single-speed machine, usually an import from the Far East. The main winding is split into two equal sections which are connected in parallel for 115V operation (Fig. 1.12a) and in series for 240V operation (Fig. 1.12b). A single 115V capacitor and capacitor start winding is provided which is connected across one of the 115V windings via the centrifugal starting switch. In 240V operation the two halves of the main winding act as an auto-transformer and provide the 115V required by the start winding.

In 240V operation the auto-transformer connection of the two halves of the main winding permits a very simple forward/reverse switching arrangement. Figure 1.12c shows the details.

1.11 Five terminations
Five terminations usually means a two-speed motor with the winding arrange-

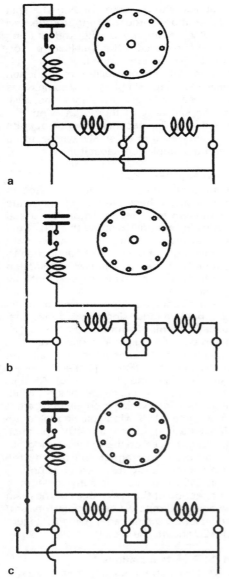

ment shown in Figure 1.13. Apart from two-speed motors used in some early types of automatic washing machines, these are quite rare. This is just as well because unless you have at least some information, such as the number and value of any associated capacitors, they are not easy to sort out.

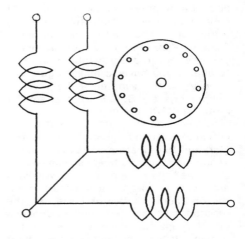

Fig. 1.13 *Two-speed motor winding*

a

b

c

Fig. 1.12 *Dual-voltage motor*

The high-speed winding will always be the highest power rating and may be capacitor start or capacitor run. In principle, split-phase start is also possible but I have never encountered one. The same possibilities exist for the low-speed winding but the split-phase variant is even more improbable.

The first thing to do is to make a careful set of measurements of the resistance between all pairs of terminations and draw up a matrix similar to Figure 1.14. A digital multimeter is almost essential because few analog meters can measure with the necessary accuracy.

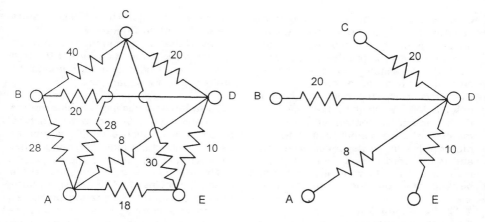

Fig. 1.14 *Resistance matrix*   **Fig. 1.15** *Predicted winding arrangement*

This matrix is not as fearsome as it looks. The lowest resistance is usually the high-speed main winding and the next lowest resistance the high-speed capacitor-fed winding: (A to D) and (E to D) are the obvious candidates. Since (A to E) = (A to D) + (E to D) then D must be the common termination which means that the remaining two points, A and B are the low-speed terminations.

The low-speed windings may fall almost anywhere in the lower resistance measurements but, as in this case, they are often symmetrical so look for two equal resistances connected to the common end of the high-speed windings. If the assumptions are right this identifies all four windings and their common point. Draw the winding arrangement corresponding to this with the four-resistance values marked in Figure 1.15 and check that this arrangement of resistance values is consistent with the measured matrix of Figure 1.14.

For appropriate capacitor values use the guidance values recommended for two- and three-terminal motors. In the case of ex-washing machine two-speed motors, 16 μF for the high speed winding and 2 μF for the low-speed winding is a good starting point.

If the resistance matrix method doesn't yield conclusive answers don't despair but simply try the spin-start method on the low-resistance windings. This should identify the high-speed main and capacitor phases.

The spin-start method allows each winding to be checked individually to see how it performs as a main winding and immediately identifies it as a high-speed or low-speed winding. Unfortunately this really needs two people – one to spin the motor up to speed and a second to apply power at the right moment. Operator one wraps a suitable length of cord several turns round the motor shaft and grasps a knot tied in the free end. A quick tug on the cord will spin the motor up to speed and if operator two applies power as soon as the cord is clear of the shaft the motor will continue to run. If a main winding is connected, the motor will run normally and reasonably quietly. If a start winding is connected the motor will sometimes run but the torque will be poor and the winding will take excessive current and heat up rapidly.

Always grasp the cord by the knot only, DON'T wrap it round your hand. If the power is applied at the wrong moment, there is a small chance that the motor will start in the reverse direction and if the cord is wrapped round your hand you can find your fingers trapped between the cord and the motor shaft!

1.12 Six terminations
By far the most common six-termination motor is the modern three-phase machine with both ends of each of the three windings brought out to terminals so that it can be connected in either star or delta configuration. It is easily recognised because it has three equal resistance separate windings. Three-phase motors are dealt with in Chapter 2.

Any of the five-termination motors can appear in six-termination form if the two wires to the single common termination are brought out separately.

1.13 Automatic washing machine motors
The earlier types of automatic washing machine were such a prolific source of useful single-phase induction motors that they warrant a section to themselves.

The first generation in the early 1960s used fully enclosed 1425 rpm split-phase motors of about ⅓ hp rating protected by a thermal cutout so that it was impossible to damage the motor by overloading the machine. The speed change between 'wash' and 'spin' was by a solenoid-controlled two-speed gearbox which formed part of the motor assembly. A motor of this type is shown in Figure 1.16. A variant used a gearbox incor-

Fig. 1.16 *Geared washing machine motor*

porating a freewheel arrangement which automatically changed from high-speed ratio to low-speed ratio when the motor was reversed.

These are excellent motors to power any of the main workshop items and have the unique advantage that the full motor power is available both in the high speed and the low speed mode. Unfortunately there are few of them left and you will be lucky to find one.

The next generation used partially enclosed two-speed, five-wire motors of about ⅓ hp rating. The usual configuration was a 4-pole 1425 rpm 'spin' winding with a 12-pole 400 rpm 'wash' winding.

These motors are easy to recognise by examining the exposed ends of the windings. The high-speed winding occupies most of the stator slot area but the auxiliary low-speed winding can be easily distinguished by its twelve-fold symmetry occupying either the very top or the very bottom of each slot. The low-speed mode uses a symmetrical capacitor-run configuration, capacitor value about 2 μF. Most machines used an unsymmetrical capacitor-run arrangement for the high-speed winding but a few used split-phase start to save the cost of the rather large (16 μF) capacitor.

These were reliable machines and, in spite of their venerable age, a steady trickle is still reaching the scrapyards or can be found as part-exchange models junked at the back of brown goods stores. Motor failure was very rare, most machines were dumped because of faulty program switches, leaking seals or just because they were getting old and shabby. If at all possible acquire the whole machine because it is then easy to sort out the motor connections at leisure and you are also sure that you have the associated capacitors. If you are taking the whole machine make sure it has the right type of motor fitted – later models use commutator machines of the type described in Chapter Three. If the back of the washing machine is still fitted, enough of the motor can be seen by tipping the unit on its side and viewing through the base.

These motors are useful for their high-speed, high-power mode. The output power in the low-speed configuration is less than ⅒ th of that of the main mode and is seldom of any practical use in the workshop.

1.14 Dishwasher motors

Dishwashers are another source of useful motors. The main motor that drives the pump for the washer jets is an open-frame two-pole 2850 rpm capacitor-run machine (Fig. 1.17). Typical capacitor value is 8 μF. Although it is smaller in size than the two-speed washing machine motors, it is usually more powerful. The higher operating speed of the two-pole machine and the absence of a second low-speed winding enables more power to be packed into a smaller frame size.

Fig. 1.17 *Dishwasher motor*

1.15 Sealed unit freezer and refrigerator motors

These are split-phase or capacitor-start motors hermetically sealed in a stout steel canister with their associated single- or two-cylinder compressor. Units from refrigerators and small freezers are mostly in the range $1/10$ hp to $1/4$ hp. The units intended for domestic air-conditioning units are higher power devices and may be as large as 1 hp. A typical unit is shown in Figure 1.18. Small units are split-phase types but the larger ones use capacitor-start motors because of their better starting torque. Basic information on the motor current or power rating is usually stamped into the casing or on some permanently attached metal label.

These motors are not fitted with centrifugal switches and the start winding is switched in and out of circuit by a special current-sensitive starter relay. The circuit arrangement is shown in Figure 1.19. When power is first applied to the motor, the main winding draws a large current because the rotor is stationary. This current closes the contacts on the starter relay which brings the start winding into circuit. As the motor runs up to speed the current drops and this allows the contacts on the relay to open and disconnect the start winding.

Starter relays are shown in Figure 1.20. The three connections to the motor are on a triangular arrangement of glass seals and the starter relay plus a Klixon thermal protector is often mounted directly on these seals. The Klixon thermal protector is a bimetal disc which snaps suddenly from convex in one direction to convex in the opposite direction when the trigger temperature is reached, breaking the power to the motor. When it cools down it snaps back to its original shape and restores power to the motor.

In normal use these compressor units are part of a hermetically sealed system containing a fixed charge of refrigerant and lubricating oil. If any part of the system is opened to the atmosphere the refrigerant escapes as gas but the oil remains in the bottom of the compressor canister. The refrigerant is unfortunately of the type that damages the ozone layer and should first be removed by one of the facilities now set up for safe disposal.

The oil used is a specially refined variety chosen to be compatible with the refrigerant. With the refrigerant removed this is no longer important and any of the common light oils can be used. Shell Vitrea 68, often used as general-purpose lathe lubricant, works fine.

Although these motors can be removed from their canisters and used as general-purpose devices, it is an operation that borders on the heroic. Sawing open the canister is bad enough but you are then faced with an open-frame motor soaked in oil and with a small compressor as a permanent part of one end plate. It may be worthwhile if you are looking for a motor of that particular size and shape, but in most cases it is better to look for a more suitable starting point!

Fig. 1.18 *Sealed compressor unit*

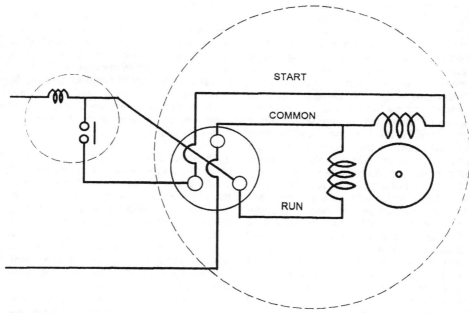

Fig. 1.19 *Compressor motor wiring*

Fig. 1.20 *Starter relays*

The main use for these units is as a vacuum pump or as an air compressor for small airbrushes. With the exception of the larger units intended for air conditioners, they do not deliver enough air to operate a full-size spray gun. Insufficient volume is the problem — the pressure that these units can generate can be embarrassingly high. On some units the pressure outlet of the compressor vents directly to the inside of the sealed canister so that this is at the maximum pressure of the system. No safety valve is needed in normal operation because the compressor is handling a fixed quantity of refrigerant which will liquefy before any dangerous pressure is reached. However once air is admitted, if the outlet is blocked, the pressure in the canister will continue to rise until either the motor stalls or the canister bursts. The limited evidence that exists indicates that the motor does not stall and the pressure builds up to over 500 psi with eventual catastrophic failure of the canister! These units must **NEVER** be used as compressors unless a reliable and effective safety valve is fitted. When charged with refrigerant the normal high-side working pressure is about 250 psi, so a safety valve setting of 150 psi should be pretty safe.

One problem with these units when used as compressors is that the air delivery contains an appreciable amount of oil mist. An effective filter is a six-inch length of two-inch pipe mounted vertically and stuffed with knitted mesh, stainless steel pot scrubbers. A drain cock will be needed at the bottom to bleed off trapped oil and water condensate. This will slowly drop the oil level in the canister so a dip stick will be needed to keep a check on this. The charge of oil in a small refrigerator unit is about 500 ml which is a layer about an inch deep at the bottom

of the canister.

The units are also useful as trouble-free quiet vacuum pumps useful for de-aerating lost-wax casting investments or as a vacuum source for solder suckers used in electronic maintenance. Because the pressure side is vented to atmosphere no safety valve is needed, but care should be taken to ensure that the outlet is not obstructed and cannot be accidentally blocked.

1.16 Shaded-pole motors

Shaded-pole motors are a special form of single-phase induction motor in which the start winding is replaced by one or more copper shorting rings which surround part of the projecting pole pieces of the motor stator. In a roughly similar way to the start winding of a split-phase motor, the currents induced in these rings distort the pulsating field generated by the main stator to generate starting torque. However, because they remain in circuit all the time and cannot be disconnected, the losses in the rings severely degrade the performance and efficiency of this type of motor.

They are mostly found in two- or four-pole form but, because of the poor efficiency, the full-load speed is only about two-thirds of synchronous speed i.e. about 2000 rpm and 1000 rpm. Full-load efficiency of the larger types is not much better than 30% and the efficiency of the very small types found in musical boxes and toys may be only 1% or 2%. Starting torque is also poor, often less than half-full load torque.

The one factor that redeems this type of motor is that it is the simplest of the induction motor family and can be manufactured at very low cost. It is particularly suitable for very small motors because it needs no additional starting components. Much of the design effort on these motors has been directed at reducing cost rather than improving performance, and this has resulted in a series of designs which are very different from the essentially cylindrical multi-winding configuration of the parent family. Some examples are shown in Figure 1.21.

The motor on the right-hand side is a four-pole machine with four salient poles projecting inwards on the stator. To reduce manufacturing cost, windings are not fitted to each pole piece but lumped into just two coils surrounding two of the four projecting poles. Each of the pole pieces is fitted with a copper shading ring which surrounds about one-third of the pole-face area.

The motor on the left is an even more radical departure. It is a two-pole machine and this time the only winding is a single bobbin-wound coil in the centre of a stack of 'U'-shaped laminations. A second stack of laminations interlocks with the first and carries the shading rings and the tunnel for the rotor.

This second type is often found as the driving motor for the centrifugal pump that empties the water in automatic and twin-tub washing machines. They are usually fitted with a fan that blows air over the motor body. Do not remove this fan because the losses in this type of motor are so high that, without this extra cooling, overheating can occur even when operating at no load.

Since there is no accessible starting winding, the direction of rotation of

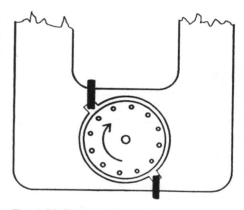

Fig. 1.22 *Shaded-pole rotor direction*

Fig. 1.21 *Shaded-pole motors*

shaded-pole motors cannot be reversed electrically and the periphery of the rotor will always rotate from main-pole area to shaded-pole area (see Fig. 1.22) However, it is easy to mechanically reverse the direction of rotation by removing the rotor, turning it end to end, and replacing it in the stator tunnel. This is usually a very simple operation involving no more than removing and replacing two screws.

CHAPTER 2

Three-Phase Induction Motors

2.1 General

There is a widespread belief that three-phase motors can only be operated from industrial 415V supplies or, in the home workshop, via expensive commercial phase converters. While there is a degree of truth in this if the motors are of the very old three-terminal variety, it is certainly not true of the now almost universally used six-terminal dual-voltage types. With some very simple additional starting gear these motors will happily operate from domestic single-phase 240V AC supplies and are often a convenient and low cost alternative to single-phase machines.

2.2 Three-phase operation

Three-phase motor construction is very similar to the single-phase types described in Chapter 1, section 1.2. The key difference is that, instead of a single main winding, it has three identical sets of windings equally spaced round the circumference of the stator. Each of the three windings is connected to one of the incoming three phases.

A three-phase motor is roughly equivalent to three single-phase motors rolled into one, driving a common rotor. Although the three winding are normally interconnected and all provided with power from the three wires of the three-phase source, it is also possible to use them individually. Provided the rotor is first spun up to speed any one of these three windings can drive the rotor as a single-phase machine.

With only a single winding energised, a single-phase pulsating magnetic field is generated and, as with the single-phase motors described in Chapter 1, section 1.3, there is no starting torque. However, when the other two phases are also energised from a three-phase supply, the three magnetic fields combine in a way which cancels the pulsations and leaves a pure rotating field. This rotating field produces torque right down to zero speed so that it is not necessary to use any of the special starting arrangements needed for single-phase motors.

The above is a drastically simplified explanation of three-phase motor operation intended only to provide a reasonable understanding of what is going on when we make the changes necessary to operate these machines from single-phase supplies. A fuller explanation can be found in the earlier book *Electric Motors* (Nexus Special Interests).

2.3 Star and delta

The three windings can be connected together in either star or delta configuration as shown in Figure 2.1.

Normal operation requires 240V across each winding. For operation from industrial three-phase supplies the windings are connected in the star configuration. Although this presents two series-connected windings to each pair of terminals, this does not require 2 × 240V = 480V but only √3 × 240V = 415V. This is because the alternating voltages developed across each of the windings do not reach their maximum and minimum values at the same time so that the total is less than the sum of the two individual voltages.

For home workshop use we haven't got 415V so we reconnect in the second, delta configuration (Figure 2.1b) which only requires 240V between terminals. It must be emphasised that, although the motor connections have been rearranged to operate at a lower voltage, the motor horsepower is unaffected. It will consume proportionally more current and deliver its full-rated power at the lower voltage when it is supplied from a 240V three-phase supply.

2.4 Single-phase operation

This is where the fun starts because we have only a single-phase supply. The first approach is to connect the supply to any pair of the delta terminals. The motor can now operate as a single-phase motor, but first it is necessary to spin the rotor up to speed because the single-phase input does not provide starting torque.

This connection utilises all of the copper in one winding and makes partial use of the other two windings. Rated full-load current is reached in the single fully utilised winding when the motor is delivering about half its rated power. The motor is still capable of delivering most of its rated power for short periods but sustained high-power operation will cause the windings to overheat.

This simple stratagem may well deliver sufficient power for occasional home workshop applications but, in most cases, the unused third terminal needs to be provided with at least a reasonable substitute for the missing third phase. Incidentally it's worth noting that a three-

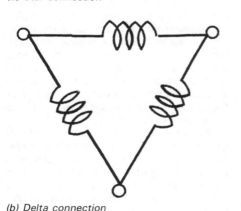

(a) Star connection

(b) Delta connection

Fig. 2.1 *Three-phase winding*

phase motor runs as a single-phase machine or as a three-phase machine — there is no intermediate two-phase mode. Two-phase operation requires the different winding arrangement described in Chapter 1, section 1.3.

A good approximation to the required voltage and phase angle can be achieved by connecting the third terminal to one side of the supply via a suitably chosen capacitor — usually called the 'run' capacitor (see Fig. 2.2). The optimum value of this capacitor depends on the hp rating of the motor, the load placed on the motor and on the detail design of the motor. There is no easy way of varying the value of this capacitor so it is usual to select a single compromise value, usually optimised for full-load operation. A list of suitable values is shown in Table 2.1.

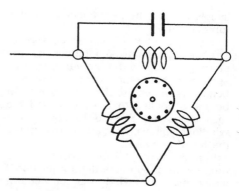

Fig. 2.2 *Run capacitor connection*

Table 2.1 'Run' capacitors

Rating hp	kW	220/240V 50 Hz (a)	(b)	(c)
0.25	0.18	10 μF	10 μF	7.5 μF
0.33	0.25	13 μF	13 μF	9 μF
0.5	0.37	20 μF	18 μF	14 μF
0.75	0.55	30 μF	25 μF	16 μF
1.0	0.75	40 μF	30 μF	20 μF
1.5	1.1	60 μF	40 μF	25 μF
2.0	1.5	80 μF	50 μF	30 μF

The values listed under column (a) are suitable for the relatively large older type motors operated somewhere near full load. The run capacitors used in the conversions described in Chapter 9 all use the values specified in this column. They are maximum values and should not be exceeded.

More modern externally finned fan-cooled motors (e.g. Figure 2.5) are much smaller for a given rating. These operate at a higher flux density, and need rather less capacity — particularly in the larger sizes. The values shown in column (b) are appropriate for operation near full load, the values in column (c) are optimum for operation at light loads.

A bonus that accrues from using a run capacitor is that the motor noticeably runs more smoothly and quietly. This is because it is operating under almost true three-phase conditions and the torque pulsations that occur with single phase excitation are much reduced.

2.5 Starting arrangements

So far it has been assumed that some unspecified method has been used to bring the motor up to running speed because single-phase operation does not generate significant starting torque. This is not strictly true because, if the full Table 2.1 value of the run capacitor is used, this usually generates enough starting torque to run a very lightly loaded motor up to speed. If, under start-up conditions, the motor is completely isolated from the load by a clutch or a loose belt it is possible to dispense with any special starting arrangements and simply rely on the permanently wired-in run capacitor.

If you adopt this solution be sure to arrange a mechanical interlock on the starting switch so that power cannot be applied to the motor unless the clutch or belt is first moved to the 'off' position. If power is applied to the motor and the rotor does not run up to speed, the windings will overheat and burn out in a matter of minutes!

In most cases the run capacitor does not provide enough starting torque and it is necessary to switch a larger value of capacitor into circuit to increase the torque while the motor is starting and running up to speed. What this capacitor does is to lower the speed at which maximum torque is generated. With a very large capacitor (about 5 times the 'run' value) the maximum occurs near zero speed. Two to three times the full load run value puts maximum torque at about half speed. This produces the maximum average torque over the run-up speed range and is generally the most useful choice. Larger values should only be used on difficult loads such as compressors which may need maximum torque as soon as the motor starts to turn.

The simple additional 'start' capacitor system cannot produce balanced three-phase drive during the run-up period, so the average starting torque is not as good as the original machine operated from three-phase supplies. Nevertheless, apart from air compressors (see Chapter 9, section 9.5), almost all workshop machines start on light load so this is not a problem.

Starting arrangements are needed to switch the starting capacitor into circuit while the motor runs up to speed and then automatically disconnect it. Chapter 2 in *Electric Motors* describes a range of different starting circuits which is com-

mended to the experimentally minded. However, in this book, it is assumed that you are not too bothered about the technical niceties of the solution but want the quickest and simplest method of powering up your new acquisition. Two cases are considered which, between them, should cover the majority of home workshop requirements.

The first assumes that a complete machine has been acquired and it is desirable to use as much as possible of the three-phase control gear already fitted to the machine. The second details the work needed if only the motor is available.

2.7 Complete machine set up

The first job is to remove the cover plate from the motor terminals and check the terminal arrangement. It will usually be a six-terminal dual-voltage motor with the terminals linked in the order shown in Figure 2.3. This is the star connection for 415V working. Move the links to the positions shown in Figure 2.4. This is the delta configuration and the motor is now suitable for 240V operation.

If it is very old machine, it may be a 415V single-voltage, three-terminal motor. If possible, replace it by a modern six-terminal motor. These are easily recognised without disassembly because they are almost invariably of the totally enclosed externally finned variety (see

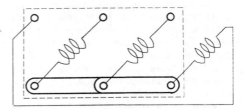

Fig. 2.3 *Three-phase terminal block — star connected*

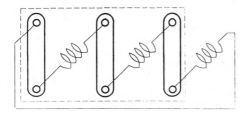

Fig. 2.4 *Three-phase terminal block – delta connected*

Fig. 2.5) and carry a nameplate which says something helpful like 400V – 440V/220V – 260V. If no suitable motor can be found convert the existing motor by the method set out in Appendix 1. This is not a job for the faint-hearted but, if carried out carefully, is usually successful.

With the links in the delta position now reconnect the three motor wires. The wires will probably not be colour coded so it is easy to forget which wire goes to where. The good news is that it doesn't matter. As long as each motor terminal has a wire connected to it the motor will function. The only uncertainty is the direction of rotation. If this is wrong, it is easily corrected at a later stage by interchanging any pair of wires.

Fig. 2.5 *Six-terminal, three-phase motor*

The three wires from the motor can now be traced back to the three heavy-duty contacts on the starting contactor inside the push button starter. These are usually marked A B C. Immediately opposite are the three-phase input terminals usually marked L1 L2 L3.

Disconnect the motor wires from A B C – they will be connected later to the new phase converter box.

Look around the contactor and the inside of the casing for useful information – the contactor coil operating voltage and the thermal trip operating range are usually displayed.

The coil voltage will usually be 415V although some special-purpose contactors are fitted with 240V coils. If there is no indication of coil voltage check the coil resistance. A 415V coil will be about 1000 ohms to 2000 ohms, a 240V coil about 250 ohms to 500 ohms.

Several methods can be used to persuade a 415V contactor to operate from 240V. Quite the best method is to fit a new 240V coil. These are available as replacement items and are stocked by the better electrical supplies factors (check your local *Yellow Pages*). However, if your contactor is an obsolete or uncommon type, replacement coils may not be available.

It is possible to reduce the operating voltage by removing about one-third of the turns but most modern contactors use resin-impregnated coils and this makes it a messy and difficult business.

The most straightforward method is to use an autotransformer to step up the 240V to 415V. This can be quite a small item as the coil is a low-power device and only consumes about 50mA. Small 240V to 415V transformers or autotransformers are quite rare, but fortunately the coil voltage is not at all critical provided it is at least 400V. Common items in electronic equipment applications are

240V to 240V isolation transformers and one of these (Maplin part No DH50E is suitable) can be used with the secondary voltage added to the primary voltage (i.e. connected as an autotransformer) to deliver 480V. It is essential that the primary and secondary connections are in the right sequence to deliver 480V — if either the primary or secondary connections are reverse connected the transformer will not be damaged but, instead of adding, the two winding voltages will cancel out and deliver zero volts!

When wired to the starter, the centre-tap of the autotransformer is connected to neutral so the full 480V only appears across the operating coil; the maximum voltage from either end to ground is only 240V. Nevertheless, don't forget that you are dealing with high voltages here — be very sure that the mains power is disconnected before touching any possibly live parts.

Do not be tempted to add a resistor to drop the output to 415V. The initial current taken by the contactor coil when it starts to operate is very much higher than the final holding current and it will fail to operate if series resistance is added. For a 415V coil, 480V is well within the acceptable range and, in fact, some coils of this type are rated for 415V and 550V operation.

The comments above refer to industrial grade contactor coils which have an ample margin of safety under normal operating conditions. The comments may not be true for less generously designed items. To be sure, the contactor should be run operated from the step-up transformer for at least an hour and a careful watch kept for signs of overheating. The coils I have checked run no warmer than hand-hot after this treatment.

An alternative source of a suitable transformer is the low-voltage bench light systems used on some of the older industrial machines. These incorporate a small transformer to reduce the mains voltage to twelve or twenty-four volts to feed a low-voltage, high-intensity bench light. Some (but not all) of these transformers are provided with 0V, 240V and 415V connections on the transformer primary to enable the lamp to operate from a 240V single-phase supply or from the 415V line-to-line voltage of a three-phase system. If a lamp transformer of this type is operated on 240V single phase there is ample spare power available from the 0V and 415V connections to operate a 415V contactor coil in addition to feeding the normal lamp load.

With the coil voltage sorted out the thermal trip is the next item. All induction motors, both single phase and three phase, take large starting currents when they are first switched on, perhaps three to six times full-load current. Fuses or circuit breakers in the mains input have to pass these large current peaks without rupturing and, while they protect the wiring against short circuits or extremely large overloads, they cannot protect the motor against a comparatively small sustained overload. The motor will happily accept short starting surges but a long-term overload of as little as 1½ times full-load current will eventually burn out the windings. Motor starting contactors are provided with thermal trips to solve this problem.

On the contactor these take the form of three bare wire heating coils which surround bimetal strips which bear against a trip bar. Because of the mass of metal involved it takes five or ten minutes for the bimetal strip to reach its final temperature. The short starting surge is ignored but, after a sustained period in excess of the rated current, the bimetal strip heats up and this makes it bend far enough to operate the trip bar. This opens the contactor by breaking the circuit to the contactor coil.

Because we have reconnected the motor in delta to operate at 240V the motor will take roughly twice as much current (1.732 ×) per phase. This is now supplied by two wires instead of three so the maximum safe supply current is now a bit more than twice the original current per phase. Most three-phase motor contactors have a calibrated adjuster which can be used to set the thermal trip current to any desired value over about a 2:1 range but this is not enough. Fortunately there is an easy solution — we simply connect all three sets of contacts (L1/2/3 & A/B/C) in parallel and wire the triplet in the live side of the 240V input. This trebles the calibration of the thermal trip adjuster and brings the motor current within its setting range.

A popular starter/contactor found on many machines is the MEM Auto Memota (Fig. 2.6). When the front cover is removed the contact arrangement is easily visible and accessible (Fig. 2.7). It will normally be connected for 415V three-phase operation as shown in Figure 2.8a and b with the 415V coil-operating voltage obtained from wire (a) connected to the L3 terminal and a link connecting the trip contact to L1.

Fig. 2.6 *MEM Auto Memota starter*

Fig. 2.7 *MEM Auto Memota starter*

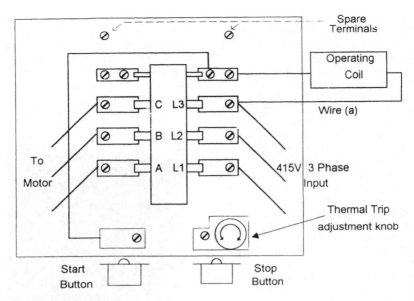

Fig. 2.8a *415V connections*

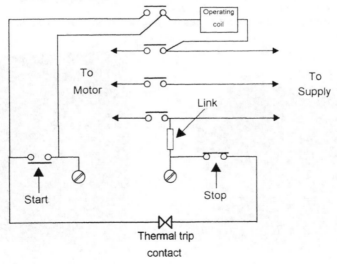

Fig. 2.8b *415V wiring*

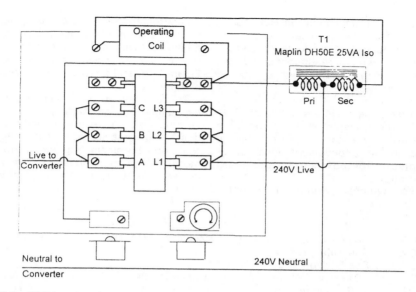

Fig. 2.9a *240V connections*

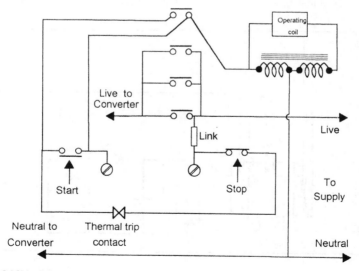

Fig. 2.9b *240V wiring*

33

The coil circuit should be isolated by disconnecting the wire (a) from L3. At the extreme back of the contact assembly there are two spare unconnected terminals. Reconnect this wire to one of these terminals.

Now rewire the starter/contactor to the arrangement shown in Figure 2.9a and b. The transformer T1 is too large to fit inside the starter case but can be fitted in any convenient location or mounted in the box housing the phase converter.

These two changes convert the motor contactor to 240V operation and the next job is to manufacture the phase converter. This is a very simple unit — the circuit arrangement is shown in Figure 2.10 and a typical completed unit in Figure 2.11.

This unit houses both the start and run capacitors and the starting circuits. The starting system uses a relay which leaves a starting capacitor connected across the motor terminals long enough for the motor to run up to speed and then disconnects it ready for the next starting cycle.

C1	10 μF 450V	Maplin 10PC450V	JL11M
C2	See Table 2.3		
C3	See Section 2.5		
R1	10 kohms 3W	Maplin W10k	W/W Min
D1	400V bridge rectifier	Maplin W04	QL40T
RL1	240V 2P C/O 10A	Maplin 240VACDPDT	JG60Q
	Relay socket	Maplin SKT 10A 8P	JG54J
	Terminal block	Maplin 15A	HL54J

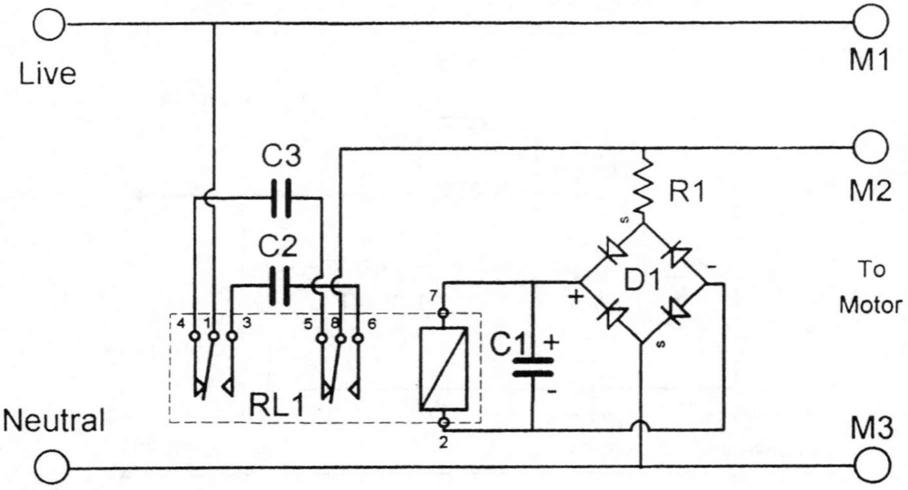

Fig. 2.10 *Phase converter*

34

Fig. 2.11 *Phase converter*

When power is first applied, C1 is initially completely discharged and the start capacitor C3 is connected across the M1 and M2 terminals of the motor. The voltage between M2 and M3 is low because of the heavy current taken by the stationary motor. As the motor runs up to speed the voltage across M1 and M2 rises and C1 charges up through R1 and D1. When the voltage across C1 reaches about 90V RL1 operates. This disconnects the start capacitor C3 and replaces it by the run capacitor. The motor continues to run with the relay operated and the run capacitor in circuit.

When the motor is switched off, C1 discharges through the relay coil ready for the next start cycle.

DC operation is chosen for RL1 to make it possible to use a standard relay at a lower operating voltage. Although RL1 is nominally a 240V relay, this is its rating when operated from an AC supply. When, as in this circuit, the AC is rectified to DC, the operating current level is reached at a much lower voltage. R1 is included to adjust the AC cut-in voltage to about 200V.

Although, at first sight, it would seem better to switch the start capacitor directly in parallel with the run capacitor, this circuit uses RL1 to select one or the other and avoids the parallel connection. This is to avoid the very high-peak currents which would otherwise occur when an initially uncharged capacitor is connected to a second capacitor which may already be charged up to the peak value of the mains voltage.

Information on suitable capacitor types is given in Chapter 8.

The circuit values are chosen for 240V operation using a start capacitor C3 two to three times the value listed in column

(a) of Table 2.1 and a run capacitor C2 from column (b) or (c).

These values are suitable for the great majority of small workshop applications e.g. lathes, mills, shapers, circular saws, drill presses etc.

If operating from 220V supplies, or using a less sensitive type of relay, it may be necessary to adjust the value of R1. The easiest way of doing this is to add a second resistor in parallel with R1 to lower the cut-in voltage — 20 kohms parallel connected will be about right for 220V operation

If a large start capacitor (four to five times Table 2.1 values) is used to increase the starting torque, the voltage on the capacitor-fed phase will rise very rapidly with speed and RL1 may cut-in and disconnect the start capacitor too soon. A small increase in the value of R1 can cure this but the adjustment will be fairly critical. A better solution is to delay the cut-in by a large increase in the value of C1 — 100 μF is a good starting point.

Construction is straightforward. The terminal block and the relay mounting base provide convenient termination points for C1, R1 and D1. R1 gets fairly warm so it should not be in direct contact with PVC wiring. C1 is an electrolytic capacitor and must be connected the right way round — the negative wire is indicated by a white stripe with - - - markings. Make sure that the unit is properly enclosed in an earthed metal case and that all live parts are safely protected against cutting fluids and accidental contact.

With the phase converter completed, connect the original three motor wires to the converter output. Connect the converter input to the switched 240V live and neutral outputs from the modified starter contactor. Check all connections and make sure that any exposed metal parts are safely earthed.

The power input to the starter contactor for motors up to 1½ hp rating can be from the normal domestic ring main via a standard 13A fused plug. Larger motors should be supplied from their own permanently wired fuse box. Appropriate ratings are shown in Table 2.2.

Table 2.2
Fuse ratings — single phase operation

Motor Power	Fuse rating
Up to 1 hp/0.75 kW	13 A
1½ hp/1.1kW	20 A
2 hp/1.5 kW	20 A

These fuses protect only the wiring to the motor. Long-term lesser overload protection is provided by the thermal trips in the starter/contactor. *Workshop Electrics* (No. 22 in the Nexus Special Interests Workshop Practice Series) gives good advice on the installation of suitable workshop power points.

With the motor electrics safely in place you should now have a working installation which will behave in much the same way as the original three-phase set up. But there is one important point of difference that should be remembered. If a three-phase motor is so heavily overloaded that it comes to a standstill, the motor will restart as soon as the overload is removed. Under some conditions, a three-phase motor phase converted by these methods will not restart but remain at standstill. The stop button must be pressed to remove power and reset the starting circuits.

Motor sizes are normally chosen to be large enough in relation to the machine size, so that this can never happen. Single- or three-phase motors should **NEVER** be overloaded to a standstill.

However, if some accidental foulup should jam up the works it is important to remove power before the motor overheats.

2.6 Motor only conversion

This is basically identical to the complete machine set up, but with the starter/contactor replaced by a simple on/off switch or by push buttons operating a latching relay.

Both systems have the disadvantage that there is no thermal cut-out to give protection against long-term small over-loads. They should not be used for un-attended equipment operation as a worst case overload fault could result in burnt-out windings. For this sort of operation a thermal push-button starter or a starter/contactor should be used.

Simple on/off switching is normally acceptable for the smaller workshop machines but lacks a 'no volt release' facility i.e if the machine is left switched on during a power failure it will start without warning when power is restored.

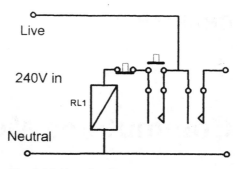

Fig. 2.12 *No volt release*

This facility should always be provided on larger or particularly hazardous machines such as circular saws.

Figure 2.12 shows a latching relay arrangement which both provides push-button starting and 'no volt release' protection. This is a useful arrangement for small motors up to about ½ hp, but for larger motors it is better to use a standard starter/contactor.

CHAPTER 3

Commutator Motors

3.1 General

Commutator motors are an entirely different type of motor. This time the main winding is on the rotor (now called an armature). It rotates within a fixed magnetic field provided, either by a set of windings on the stator (now called the field) or by one or more permanent magnets. The armature and field assembly of a commutator motor is shown in Figure 3.1.

The armature winding is tapped at many points and these points are connected to a rotating switch (the commutator). The commutator consists of a circular array of copper bars that rotate with the armature. Two or more fixed contacts (usually carbon brushes) bear on this commutator. As the armature rotates, this automatically switches the current flow to different points in the windings so that the interaction between these currents and the fixed field always produces a torque

Fig. 3.1 *Commutator motor components*

which tries to rotate the armature in the same direction.

A commutator motor can act both as a motor and as a generator. If the armature is rotated, it will generate a voltage and this will appear across the brushes. This voltage, called the Back Electromotive Force (back EMF), is directly proportional to the speed, the fixed field strength and the number of turns in the armature winding.

If a voltage is applied to the brushes, current will flow and the torque that this produces will start to rotate the armature. This rotation of the armature will generate a back EMF which will oppose the applied voltage. In a perfect motor, with no mechanical or electrical losses, the speed will rise until the back EMF equals the applied voltage. Because the two voltages are now equal, no current can flow and the armature will then be rotating at a speed directly proportional to the applied voltage.

In a real motor there will be mechanical and electrical losses. This causes the armature to slow down slightly so that the back EMF is no longer equal to the applied voltage.

If a mechanical load is now placed on the shaft, the armature will slow down a little more until enough current flows to generate the torque necessary to turn the shaft.

Commutator motors come in four main types:

Permanent-magnet motors
Shunt-wound motors
Series-wound motors
Compound-wound motors

For a particular supply voltage, shaft speed and output power all four types could use the same armature. The difference lies in the field arrangements.

3.2 Permanent-magnet motors
These come in a wide range of sizes but are particularly popular in the smaller sizes. They are the most efficient type of commutator motor because no power is required to maintain the permanent magnet field. A range of PM motors is shown in Figure 3.2.

Fig. 3.2 *Permanent-magnet motors*

This type of motor must run from a DC or a rectified AC supply because the direction of shaft rotation is controlled by the polarity of the supply. The direction of rotation can be reversed by interchanging the connections to the armature.

In an ideal PM motor the speed is directly proportional to supply voltage and the current drawn from the supply is directly proportional to the torque load on the shaft. The larger PM motors are over 80% efficient so their performance is reasonably close to that of a perfect machine. The shaft speed of a PM motor can be varied over a wide range by control of supply voltage and, at any set voltage, the speed only drops slightly as the load on the shaft increases.

This is very different from the induction motors that we have looked at so far. Induction motors are designed to operate at a fixed speed from a specific supply voltage and frequency, and these operating conditions cannot easily be varied.

It is perhaps obvious that a PM motor can be operated at a reduced voltage if lower speed and power output are acceptable. What is less obvious is that, in some cases, it is equally permissible to operate it at a higher than normal supply voltage and deliver more power than its original rating.

The important rating in a PM motor is the armature current. At low and medium shaft speeds the major power loss in the motor is the result of this current flowing through the resistance of the armature windings. The supply voltage controls the shaft speed but has a much smaller effect on the power dissipated within the motor. Provided the rated maximum armature current is not exceeded, the supply voltage can be varied over a wide range both above and below nameplate voltage to suit the desired operating conditions. The nameplate operating voltage and shaft speed are only one pair of a wide range of equally valid shaft speed and operating voltage pairs.

These comments on higher voltage operation mainly apply to low voltage motors that operate at a modest speed. A good example is the 12V 4A fan motor fitted to many car heaters. These run at about 5,000 rpm simply because, at any higher speed, the fan would be too noisy. With the fan removed, they will run at 18V and deliver about 1½ times their normally rated power. (Do not leave the fan on – at 1½ times the rated speed a fan absorbs more than three times as much power and that will certainly overload the motor!)

Mains-voltage or high-speed motors are not generally suitable for running much above their nameplate voltage. Mains-voltage motors may suffer from excessive commutator sparking or even complete flashover. High-speed motor designs are usually the result of an attempt to achieve maximum power at minimum size and cost. The distribution of losses is different and any increase in voltage is likely to do more harm than good.

Care should be taken when running any motor above its rated speed especially in the larger sizes. The centrifugal forces on the armature conductors increase four times when the speed is doubled so there is a danger that the windings may break loose. The bindings and the resin impregnation systems used in most motors are fairly rugged and this is unlikely to be a problem at modest speeds. Nevertheless be very sure to choose your working conditions so that if the armature does fail the results are not a danger to you or anyone else.

Apart from a few specialised types, PM motors use ferrite permanent magnets. This is a black ceramic-like material which can operate at a high enough magnetic flux density for most motor applications and has the great advantage that only a short length of magnet is required. A common form of construction uses a soft iron tube which forms the outer casing of the motor. The magnetic material takes the form of two radially magnetised arcs which are bonded to the inside periphery of the cylinder.

For most purposes these magnets can be regarded as truly permanent. They do not wear out and are not degraded by the normal use of the motor. However some specialised types of motor use a different type of magnet material and, in some circumstances, this can become partially demagnetised. In extreme cases this can also occur with ferrite magnets. For more information see Appendix 3.

3.3 Shunt-wound motors

In these motors the permanent magnet field is replaced by an electromagnet usually composed of two saddle-shaped coils fitted to a stack of laminations that form a two-pole field assembly (see Fig. 3.1). This type of motor cannot be used on AC supplies so the special characteristics of laminated iron circuit are not essential. Larger motors often use solid iron yokes with bolted on pole pieces. Occasionally a four-pole field assembly is used. In an induction motor the number of poles in the stator controls the shaft speed of the motor. In a commutator motor it has no major effect and the number of poles is mainly chosen as a matter of manufacturing convenience. Any of the commutator motor types may use a multi-pole field, but apart from noting that the angle between the brush holders is changed (to 90 degrees with a four-pole field) it makes no practical difference to the user.

The shunt-field winding is designed to be connected directly across the armature supply and will typically absorb 5% to 10% of the input power. If the shunt field is connected to a separate fixed supply the motor will behave in exactly the same way as a PM motor. This connection is frequently used if the application needs the speed to be varied over a wide range by control of armature voltage.

A shunt-field motor can be reversed by changing the polarity of the armature voltage *or* the field voltage. It is occasionally useful to be able to reverse a shunt motor by changing both i.e. reversing the supply voltage. This is possible with the arrangement shown in Figure 3.3. In this case the bridge rectifier ensures that the field current is always in the same direction and the change in supply polarity only affects the armature.

With the shunt field in its normal connection, directly across the armature, the speed change that would normally accompany a change in armature voltage is almost cancelled out by a similar change in field strength. For a small voltage range, about the design operating voltage of the motor, the speed is almost constant. The range in the upward direction is limited because of non-linearities in the iron circuit and also by overheating of the field windings. In the downward direc-

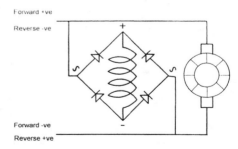

Fig. 3.3 *Shunt motor reversal*

tion the efficiency of the motor worsens as the field strength reduces with consequent reduction in speed.

Nevertheless, a shunt-wound DC motor operated from a fairly constant voltage source is a good approximation to constant speed motor and very suitable for the type of application that would normally use an induction motor.

The constant speed can be varied over a small range in the upward direction by putting a variable resistor in series with the field winding. The efficiency and speed regulation get rapidly worse as the field is weakened so this is only really suitable for trimming the motor to a particular speed rather than changing the speed by any large amount. Speed can be reduced by putting a variable resistor in series with the armature but the results are even less satisfactory. Large amounts of power are dissipated in the resistor and the speed is then very dependent on the mechanical load.

3.4 Series-wound motors

The construction of series-wound motors is almost identical to the shunt-wound variety but with the difference that the field winding is now made up of a comparatively few turns of thick wire. Because it is connected in series with the armature (Fig. 3.4a) the field current is now always equal to the armature current. This increase in field current with increase in armature current results in a drooping speed/torque characteristic i.e. the speed drops as the load on the motor increases.

In a perfect loss-free series-wound motor, operating from a fixed voltage, the shaft speed halves each time the torque load on the motor is quadrupled. If no load at all is placed on the shaft, the speed would rise without limit. In a real motor, resistive losses in the armature and field result in the speed dropping more rapidly as the load is increased. Air resistance and increasing iron losses limit the no load speed.

This speed/torque characteristic is useful for some vehicle traction requirements but in the workshop the constant speed characteristic of the shunt-wound motor is usually preferred. The inherent tendency of series-wound motors to overspeed can be dangerous and operation without shaft load should be avoided. Many of the smaller types can withstand at least short periods of no-load operation, but some of the larger or higher speed types

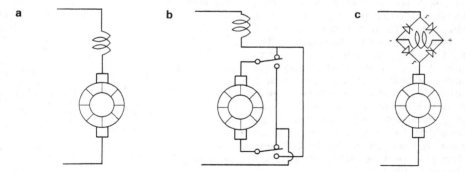

Fig. 3.4 *Series motor connections*

can reach speeds high enough to cause conductors to break loose from the armature.

Series-wound motors can be reversed by the same methods as shunt-wound motors, Figures 3.4b and 3.4c show the details. Split-series motors (Fig. 3.5) are often used in applications that require frequent reversal. These use a separate field winding for each direction of rotation to simplify the switching arrangements.

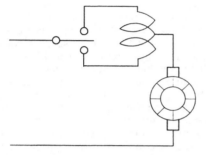

Fig. 3.5 *Split-series motor*

Reversing works fine on motors operating at speeds of up to a few thousand rpm because the brushes are in the symmetrical neutral position and the direction of rotation doesn't matter. However, many series-wound motors are high-speed machines and these have the brush assemblies rotated slightly away from the neutral position to improve commutation (i.e. to reduce sparking). These motors can still be reversed but there will then be a some loss of efficiency and increased sparking at the commutator.

The key advantage of the series-wound motor is that it is that it is the only type of commutator motor that can work directly from AC mains supplies and, because of this, it is more widely used than any other commutator motor type. It can operate on AC because the series connection ensures that the current in the armature reverses at exactly the same time as the current in the field windings so that the torque is always maintained in the same direction.

A shunt-wound motor does not behave in the same way. The shunt field has a high inductance and when operated from an AC supply this inductance both reduces the field current and delays the time at which it reverses. The reduced field current drastically reduces the torque that the motor can generate but this is only part of the story. The current in the field does not reverse until sometime after the current in the armature has reversed. This means that for part of each cycle of the supply frequency the torque generated is in the wrong direction! This results in the net output ranging from nothing to a tiny fraction of its DC performance.

The field winding of a series-wound motor also acts as an inductance but, because of the reduced number of turns, it is much smaller in value and has only a minor effect on motor operation. This inductance behaves as a sort of AC resistance in series with the motor which slightly reduces the full-load speed and power. It also has the useful effect of limiting the maximum current that can flow if the motor is severely overloaded.

AC operated series-wound motors (often called 'universal motors') are used in portable power devices such as vacuum cleaners, electric drills and power saws. They are also used in fixed domestic devices such as washing machines, spin driers and food processors. The portable devices are often fitted with simple speed control systems based on a triac power semiconductor and a few associated components. These work mainly by controlling the average value of the input voltage to the motor. The output speed still drops as the load on the motor increases and the speed range is fairly limited, but the additional cost is small and the performance is generally good enough for the intended application. The circuit arrangements used are very similar to the electronic dimmers used in domestic lighting and some of these dimmers are suitable for speed control of small series motors.

Automatic washing machines use a more complex control system to achieve tight speed control over a wide speed range. This is a true closed-loop speed control system and is discussed in Chapter 5.

3.5 Compound-wound motors

Compound-wound motors carry both shunt and series field winding. They are basically shunt-wound motors with a small additional series field to modify the starting characteristics. They are only encountered in the larger sizes and the terminal block often uses the coding shown in Figure 3.6.

In large motors the shunt-field current may take a second or so to build up to its final value. For much of this time the

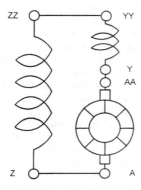

Fig. 3.6 *Compound-wound motor connections*

armature is stationary because, with insufficient field current, it cannot develop enough torque to turn the load. A stationary armature draws a very large current from the supply. This may cause such a large voltage drop in the supply leads that the shunt-field current can never reach the value needed to turn the load and the motor will fail to start.

In a compound-wound motor, the series field turns only contribute a small part of the field flux — perhaps 5% or 10% at normal full-load current — so the motor behaves almost as a pure shunt-wound machine. Now, at start up, flux from the series field is available as soon as current starts to flow and with a 10% series field, full-load torque is available when the initial starting current reaches about three times normal full-load current (torque is proportional to the product of field current and armature current). This torque is current controlled and is still available even if the heavy current has dropped the supply voltage well below normal.

The magnetic flux from the series winding must always aid the shunt-winding flux and this must be maintained if the motor is reversed. Motors can be reversed by reversing the connections to the armature or by reversing the connections to *both* the series- and the shunt-field windings.

If it is necessary to operate a straight-shunt machine on long supply leads, the starting problem can be eased by using a four-wire cable — two wires for the field and two for the armature. There is still a delay while the field builds up to full strength but the large armature currents that flow no longer cause additional voltage drop in the field supply cables and the full field strength is available. Since the field current is much smaller than the full-load armature current, thinner cables can be used for the field supply — typically ⅕ to ¹⁄₁₀ of the copper cross section of the main armature cables.

3.6 Common types of commutator motor

Commutator motors are manufactured in an enormous range of sizes and types but, for home workshop use, motors from domestic machinery and automobiles are the most readily available and with a little ingenuity they can be used in a wide range of workshop applications.

Domestic washing machines are a fruitful source of useful motors. In their original application they are reasonably well protected against overloading and misuse and most are still fully functional when the machine is discarded.

Figure 3.7 shows two ex-automatic washing machine motors. The smaller one on the left is from one of the first of the automatics to change from induction motors to series-wound commutator motors in the quest of more power at less cost. The larger motor on the right is from a later model fitted with a higher

power motor to achieve 1,000 rpm spin speed. Variants of these types of motor are fitted to different makes of washing machine but these two frame sizes are representative of the common types. Normal full-load speed of both types is about 8,000 rpm. The smaller motor can deliver about ⅔ hp and the larger about 1 hp. Both are fitted with a very small eight-pole permanent magnet generator mounted on the end plate at the commutator end of the motor. The frequency of the AC output of this generator is converted to a voltage by a frequency to voltage converter in the electronic control system and used to monitor the shaft speed of the motor.

These motors are obvious candidates to replace the induction motors normally used to drive small lathes, milling machines and pillar drills. They have ample power and the possibility of varying the speed is a useful plus. However, a fair amount of work is needed to turn them into useful workshop workhorses. If a quick and simple method of powering a machine is

Fig. 3.7 *Automatic washing machine motors*

45

needed, and you are not looking for variable speed, then an induction motor is a better bet. One point that should not be overlooked in the home environment is that an induction motor at a leisurely 1,425 rpm. is a great deal quieter than a commutator motor screaming away at 8,000 rpm.

Two factors usually determine the choice. First, if you really need variable speed then this is the best way to go. The second, and usually overriding, reason is that if you've already got the motor it seems a pity to throw it away!

The first thing to do is to sort out the connections. These are unfortunately are not standardised but the connections are quite easy to identify — the following notes should help.

1. There may be up to 9 connections

	No of connections	Typical resistance
Armature	2	2 ohms (see note 2)
Series field	2	2 ohms
AC generator	2	180 ohms
Overtemperature	2	(see note 3)
Motor frame (earth)	1	—

2. Armature connections are easily distinguished from the similar resistance series field by lifting one brush. This will make the armature connections read open circuit but make no difference to the resistance measured at the field connections. The true armature resistance measured between diametrically opposite bars of the commutator will be about 2 ohms. The apparent armature resistance measured with an ohmmeter at the input terminals may be considerably higher — as much as 10 ohms because of the non-

linear contact resistance of the carbon brushes. This high brush contact resistance only appears because the ohmmeter measures the resistance at a very low voltage and current. At normal operating voltage and current the resistance will be much lower and the voltage drop across each brush contact will be less than a volt.

3. There may be no overtemperature sensor. If one is fitted it may be a self-resetting bimetal thermostat or a thermally sensitive resistor for use by the electronic control system. The bimetal thermostat will normally read zero ohms. Posistors used for thermal sensing read an ohm or so at room temperature but suddenly rise to a much higher value at their trigger temperature.

The next step is to check that the motor runs. Secure the motor to the bench and earth the frame. These are large high-speed series motors and should **NOT** be run unloaded at full voltage. Even if the armature remains intact, the bearings and frame are not designed for operation at these speeds (over 20,000 rpm!) and you are likely to encounter early bearing failure or fatigue cracks in the comparatively light frame.

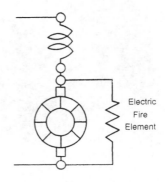

Fig. 3.8 *Series motor test rig*

46

Connect the motor as in Figure 3.8 with a 750W or 1,000W electric fire element connected directly across the brushes. The current flowing through the fire element passes sufficient current through the field windings to limit the no-load speed to a safe value – this arrangement has the same effect as adding a small shunt-wound field winding to the normal series field. The motor should run at close to its normal operating speed with little or no sparking at the brushes. A little sparking is permissible but sparks that leap across one or more commutator segments are sure indications of a faulty armature – fortunately a fairly rare event.

The electric fire element shunt is a useful dodge for checking these motors out on the bench but it's rather wasteful of power. These motors are designed to operate under electronic speed control and, to make good use of them, they need at least one of the simpler forms of electronic control. Suitable speed controllers are described in Chapter 5.

If the motor is to deliver its full power it must be allowed to reach about 8,000 rpm. They're fairly noisy at this speed and for the sake of a peaceful life you may settle for half power and about 4,000 rpm. Because many machine tools can leave the motor running with almost no load on it, it is not advisable to run the motor at full mains voltage unless it is fitted with some form of speed controller.

If it is replacing an induction motor, at least one additional stage of speed reduction will be needed to bring the full load output speed to the 1,000/2,000 rpm region. Vee belt reductions are not very satisfactory at high speeds and with small pulley diameters. Timing belt or poly-vee belt drives give better results. This is discussed further in Chapter 5.

Spin driers are another useful source of motors. These are rather smaller series-wound motors in the power range $\frac{1}{10}$ hp to $\frac{1}{4}$ hp. In spin driers these motors frequently operate at full voltage with no more load than a vee belt and an empty spin tub so no-load operation is quite close to their normal use. The no-load speed is about 18,000 rpm and they run at this speed without obvious signs of distress. Some types are fitted with an external cooling fan – it is important that this is retained both to prevent overheating and to limit the no-load speed.

They are useful for low-power, variable-speed workshop applications – Chapter 9 describes a fretsaw powered by one of these motors.

Vacuum cleaner motors are high-power, high-speed series-wound machines. Older types deliver about $\frac{1}{2}$ hp and more recent high power types more than 1 hp. This performance is only possible because they operate at very high speed and are very efficiently cooled by the whole airflow passing through and round the motor. Robbed of this airflow and operated at a more modest speed they become undistinguished machines operating at an awkward supply voltage – too high for batteries and inconveniently low for mains operation. They lead a hard life and motors in discarded machines are rarely in good condition.

Portable power appliances are almost equally unpromising. Their shaft speed and cooling requirements are more modest but, once again, the most frequent cause of early retirement is motor failure. In addition, the motor is often built as an integral part of the appliance casing and operation as an independent unit is hardly practicable

Occasionally it may be desirable to extract a little extra power from these smaller 240V AC series-wound machines. This can be done by operating them from a rectified supply instead of directly from the mains (see Fig. 3.9). The rectifier tries to charge C1 to the peak value of the supply voltage. If no current is drawn C1 will reach about 330V. If driving a motor load this will drop to perhaps 280V to 300V but still well above the normal 240V AC input. Even if C1 is omitted there will still be a significant increase in power because, when supplied from a DC source, the inductance of the field coils no longer reduces the voltage reaching the armature. This is, of course overrunning the motor — usually OK for intermittent operation but watch out for overspeeding or overheating.

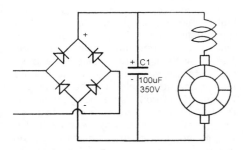

Fig. 3.9 *240V motor boost*

Motor vehicles are a good, but often expensive, source of motors suitable for battery operation. The problem here is that the seller assumes that the motor is needed as a replacement part for a treasured vehicle and even the most grotty sample tends to be sold at about half the price of brand new spare.

Window motors, sunshine roof motors and seat adjustment motors produce a lot of power for their size but are all intermittent rated. They cannot be used continuously at their full 12V to 14V rating but can produce quite useful continuous outputs when operated in the 3V to 6V range.

Continuous rated items are windscreen wiper motors and fan motors both from interior heaters and also from the main radiator. Windscreen wiper motors may be series wound or permanent magnet types and many have a two-speed facility. Older types used a tapped series field winding, later types use an alternate brush position or an electronic regulator.

There is often very little depth available for main radiator fan motors so they tend to be short, large diameter designs. Some use a very different 'pancake' motor design in which the conventional radial magnetic field is replaced by an axial one (Figure 3.10). The armature takes the form of a thin disc of conductors and does not need an iron core. In vehicle use the main advantage of this form of construction is its low cost and its very shallow depth. In the workshop it can be treated as an ordinary PM motor but the unusual shape tends to be a handicap rather than a help. Specialised types of these axial field motors are manufactured for use in high performance servo systems which can take advantage of the very low inertia of this form of armature. These are described in *Electric Motors* (Nexus Special Interests).

Car heater fan motors already mentioned in section 3.2 are useful little machines either for battery-powered equipment or run from the mains through a transformer rectifier set. The sample shown in Figure 3.11 has a speed constant of about 500 rpm per volt over the useful operating range of 6V to 20V. Safe full-load current is about 4A.

Many modern commutator motors are fully enclosed and it may not be immediately obvious whether it is a series-wound, shunt-wound or permanent-magnet machine. A simple way of discovering this is to apply a few volts to the motor terminals and observe what happens

48

Fig. 3.10 *'Pancake' motor*

Fig. 3.11 *Car heater motor*

when the supply is reversed. If the direction of rotation reverses when the supply is reversed it must be a permanent-magnet machine. If it continues to rotate in the same direction, lift one brush. A shunt-wound machine will still continue to draw its field current from the supply. A series-wound machine will draw no current at all.

A once useful, but now fairly rare, item

49

is the car dynamo. Dynamos run very happily as motors but the appropriate voltages are rather different. A 12V car dynamo designed to deliver 30A at its 'cut-in' speed of 2,000 rpm must actually generate about 16V because the normal charging voltage will be at least 13.5V and a further 2V to 3V will be lost in the armature resistance and brush drop. For it to run at the same speed and current level as a motor the field current must be maintained at its previous level, but the armature requires the 16V operating level *plus* the 2V to 3V lost in the armature resistance and brush drop i.e. 18V.

To merely duplicate its lowest 'cut-in' operating speed at least 18V is required, but this is not the end of the story. Car dynamos are rated to operate up to at least 6,000 rpm and operation as a motor at this speed needs about 50V. Somewhere between these extremes is reasonable and 12V on the field and 24V on the armature results in a constant speed motor running at about 2,800 rpm with a continuous rating of about ¾ hp.

Higher voltage operation of a dynamo as a motor is not overrunning it but simply using a different part of its speed/voltage characteristic. The same comments do not apply to the field. This must continue to operate close to its original rating – any significant increase in field voltage will lead to overheating. A small reduction to trim the speed is OK but any large reduction ruins the efficiency.

3.5 Radio interference
A point that should be remembered is that all commutator motors are prolific generators of radio interference – mainly in the medium and long wavebands. Modern equipment using commutator motors is fitted with interference suppression components usually mounted on or near the motor. Sometimes a separate packaged interference suppressor is used. Automatic washing machines use a special unit looking rather like a capacitor connected across the mains near the point at which they enter the machine. Examples are shown in Figure 3.12.

Suppression components should be retained and installed with the motor in as near their original position as possible. If the original suppression components are not available a useful reduction can be obtained by connecting a 0.01 μF capacitor directly between the brushes using the shortest possible leads. Ceramic capacitors are the best type to use. For mains-voltage motors they should be at least 500V rating (e.g. Maplin No BX15R/ 10,000 pF – 10,000 pF is exactly the same as 0.01 μF – it is just a matter of using different units to describe the same value of capacitance!). If this does not give enough reduction then add a triple capacitor motor suppressor and a pair of RF suppression chokes (see Fig. 3.13).

For larger motors, a washing machine suppressor unit installed in the incoming

Fig. 3.12 *Washing machine radio interference suppressor*

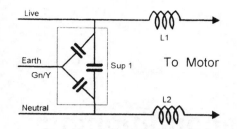

Live

L1

Earth   Sup 1   To Motor
Gn/Y

L2

Neutral

Fig. 3.13 *Radio interference suppression*

Sup 1   Maplin HW07H   Delta Cap
L1 & L2   Maplin HW06G   3A RF   SUPPRESSOR CHOKE

mains lead is usually the best solution.

In all cases the wires between the motor and the suppression units can radiate interference. To control this, these wires should either be covered with an earthed metal braid or be located entirely within an earthed metal enclosure.

CHAPTER 4

Overhaul and Reconditioning

4.1 Induction motors

Induction motors are fairly reliable devices and, if you have selected your acquisition with care, in all probability there will be nothing wrong with it. However, if you are unlucky, the likely problem areas are bearings, starter switches and start windings.

Before starting overhaul, the first thing to do is to ensure that the motor is dry. Motors can withstand being soaking wet for surprisingly long periods and mechanical failure from rust is likely to occur before the electrical insulation is permanently damaged **PROVIDED** that power is **NOT** applied to a wet or damp motor. Leakage current flowing though damp insulation can form carbon tracks which, in time, will lead to complete failure. If the motor has been wet at any time it is essential to ensure that the windings are thoroughly dried out before applying power. This can take a long time at room temperature but even a small amount of heat can accelerate the process. A fan heater or hot air gun for an hour or so are useful accelerators, but leave a reasonable gap between the heat source and any exposed windings to ensure that the air is never hot enough to damage dried out insulation. If the motor is of the totally enclosed variety it is better to remove at least one of the end bells so that air can circulate freely around the motor windings.

Once you are sure the windings are dry use the ohms range of a multimeter to check that there is no short circuit or excessive leakage between motor winding and the external metalwork. The ohmmeter should indicate infinity i.e. open circuit — anything less than many megohms shows that the insulation is faulty or not properly dried out.

With the insulation checked out, now make a direct and reliable connection between the motor metalwork and the electricity supply earth. This is essential to ensure that, when power is eventually applied, insulation failure or a misconnection cannot result in 'live' metalwork. Finally anchor it safely to the bench so that a sudden start or stop doesn't make it leap off the bench.

Use the methods described in Chapters 1 and 2 to identify the appropriate connections and apply power to the motor to check that it runs — there is no point in overhauling the mechanics of a motor that's electrically dud!

If it fails to start the most likely faults are:

All machines
Broken lead out wire
Burnt-out main winding

Single-phase machines
Faulty centrifugal starter switch
Burnt-out starter winding
Dud starting capacitor

Most of these faults are easily identified by visual inspection. The difficult item is the capacitor — usual faults are open circuit or short circuit. Partial loss of capacitance is possible but infrequent — if this is the problem it will show up as reduced starting torque, not as failure to start. Short circuits are easy to detect with a multimeter. Open circuits will appear as absence of the initial 'kick' of the needle when a multimeter switched to the ohms range is first connected to the capacitor.

With a good capacitor there should be a significant kick when the meter is first connected and the needle should drop back to zero reading, i.e. infinity ohms, as the capacitor charges up to the full voltage of the battery in the multimeter. There should be little or no kick if the meter is briefly disconnected and reconnected because the capacitor should retain its charge. If the meter is then disconnected and then, without delay, reconnected with the leads reversed a somewhat larger kick should be observed as the capacitor first discharges and then recharges in the reverse direction.

These tests are easy to carry out with an analog meter and the size of the kick gives some indication of the value of the capacitor — more μF results in a more pronounced kick. Although a digital multimeter has many compensating advan-

tages it is of little use for this sort of test because the digital display does not change fast enough to give a clear indication of the kick. Some of the more expensive digital multimeters have a facility for directly measuring capacitance but this is limited to low values of capacitance — less than 20 μF.

There is a very wide range of different designs for the centrifugal switch but the types of fault encountered are pretty standard. There is either some mechanical fault or obstruction which prevents the weights flying out and reclosing at the appropriate speeds, or the switch contacts are badly worn or excessively dirty. Major faults are rare and clean up of the contacts followed by a careful check of the mechanics will usually restore the switch to working order.

Burnt-out main windings are a different kettle of fish. Unless there is something totally unique about the motor it is just not worth the time, trouble and cost of a rewind.

A burnt-out starter winding on a split-phase motor is a marginal case. The starter winding is a small winding which only occupies a few slots and it can often be removed and new windings inserted without disturbing the main winding. If you're feeling lucky, it may be worth the effort. If you're tempted to try this, examine the main winding very carefully at any place where it has been in contact with the burnt-out winding to make sure that the insulation has not been damaged.

If the motor is electrically OK then check for excessive play in the output shaft bearing and listen for untoward noises when the motor is run up to speed.

With a motor fitted with plain sleeve bearings it should run quietly apart from a little vibration and the low-pitched hum from the 50Hz excitation. If the motor has led a long hard life there may be significant play in the output shaft bearing but sleeve bearings are pretty tolerant of reasonable wear and catastrophic failure is unlikely. Unless you are a perfectionist no action is needed. Unless there has been a lubrication failure, the bearing at the non drive end of the shaft is usually OK.

The sleeve bearings are usually sintered bronze pressed into the end housing and carry enough oil in their porous structure for many hundreds of hours of normal operation. Larger or more heavily loaded bearings may also be surrounded by an oil-soaked felt pad which acts as an additional oil reservoir. In either case a little extra oil is helpful. It is usually impossible to discover the original manufacturer's oil recommendation. For low-speed motors (less than 3,000 rpm) I use the machine oil (Vitrea 68) that also lubricates my lathe and mill. For higher speed motors I use the readily available '3 in one' oil or sewing machine oil. The important thing is to avoid the use of grease because this clogs up the pores and interferes with the intended capillary fed lubrication.

A motor fitted with ballraces needs to be treated rather differently. Ballraces are inherently noisier than sleeve bearing so a somewhat higher noise level at operating speed is not a cause for alarm. The key test is to slowly rotate the rotor by hand and feel for any roughness, grating or radial play. Ballraces tend to live a long and trouble-free life until they start to develop fatigue cracks in the surface of the inner or outer race tracks. Small scales of the track surface break away and mix with the grease to form a sort of grinding paste. Once this process has started, deterioration is rapid and the remaining useful life is at best measured in tens of hours.

If roughness or grating is detected the choice is either to scrap the motor or replace the offending bearing. Bearing replacement used to be the automatic choice but many modern motors are basically designed as throw away rather than repairable items and it may be very difficult to remove the faulty bearing without damaging the rest of the motor. Very often it is safer not to attempt to remove the bearing in one piece but to destroy it and remove it piecemeal. Grind right through the outer race at one point − it can then usually be sprung clear of the inner ring plus caged balls assembly. If it is a heavy-duty bearing it may first be necessary to first grind a deep flat or groove on the outer race, diametrically opposite the gap, to weaken the strength of the resultant 'c' spring.

Remove the exposed cage and balls and grind a pair of generous flats on opposite sides of the exposed inner race. Continue the grinding on at least one of the flats until the thinnest part of the inner is only a few thou thick. With the inner gripped in a vice by the two flats, a forceful twist will crack it open at the thinnest point and it will slide easily off the shaft.

Occasionally the dud race is easily accessible but is still very difficult to remove. This usually means that it has been secured with one of the anaerobic locking compounds. In this case the bearing will be a sliding or very light interference fit and if the assembly is heated to about 150°C, the glue will soften and the bearing can be slid off easily. This must be done while the assembly is still hot because, as soon as it cools down, the glue regains much of its strength.

Ballrace sizes are pretty standard and it should not be too difficult to obtain a replacement from a specialist bearing supplier. However it may be worth considering that you are very unlikely to need the thousands of hours of use that the correct replacement can endure. A replacement made in the form of a plain bearing mounted in a dummy of the ballrace may well be all that is needed. Oilite sintered bronze bearing sleeves are readily available in appropriate sizes and mounted in a suitable slug of mild steel or light alloy are a simple replacement for an expensive or hard to find ballrace.

If you choose to use a replacement ballrace remember that you must **NEVER** force it on to the shaft by applying pressure to the outer race. The inner race is an interference fit on the shaft and considerable force is needed install it. If this force is transmitted via the outer race and the balls, the contact pressure is high enough for the balls to make small indentations on the race tracks and this will cause early failure.

If possible use a press or a vice to provide steady controllable pressure via a sleeve sliding over the shaft and butting squarely against the race inner. A light dab of anti-scuffing paste or molybdenum disulphide grease will act as an extreme pressure lubricant and lessen the chance of a slightly misaligned bearing seizing on the shaft. This is a fairly trouble-free set up and you should have no problems.

Life is a bit more difficult if the rotor is too big to fit in the vice or the press and it is necessary to drive the bearing on to the shaft with hammer blows. A successful installation is still possible but you need to be sure that the driving sleeve is a close fit on the shaft and that you use a good heavy club hammer. A light blow with a heavy hammer is safer and more effective than a heavy blow with a light hammer.

The force required to install the bearing can be greatly reduced if it is first expanded by cooking it for twenty minutes in the domestic oven at 100°C. Don't forget to put something under the bearing to catch any oil that may leak out during the cooking process. The bearing will then readily slide onto the shaft but it is important that it is pushed quickly and without hesitation to its final position — once the bearing has made good contact with the shaft it almost immediately cools to shaft temperature and the benefits of the preheating disappear.

A fairly common bearing fault is a bearing that passes the slow rotation test with flying colours but appears to have a lot of radial play and is rather noisy when run up to speed. This is usually a good bearing that has worn a clearance between its outer ring and the end bell housing of the motor.

In initial manufacture, the fit of the bearing outer into the housing is quite critical. It cannot be a heavy interference fit or the forces required to assemble and disassemble the motor will be too large and damage the race tracks. Equally, there must not be significant clearance between the bearing and the housing or the housing will not properly support the outer ring and it will flex very slightly as the peak loads transmitted by the balls rotate round the inner circumference. Although this flexure is extremely small it causes the outer ring to slowly 'walk' round the inside of the housing. There is no lubrication at this interface and, over a long period of time, wear increases the clearance to the point where the motor becomes unacceptably noisy. If the motor is disassembled it is quite easy to spot this — even in the early stages. A blue-black discolouration develops on the bearing outer which contrasts strongly with the normal fine-ground finish.

Provided the wear is not excessive an easy and permanent cure is to secure the bearing in its housing with one of the anaerobic locking compounds. A high-strength compound is not needed. Loctite 270 is formulated for this sort of application and is easy to use. If the wear is severe (>0.010") it approaches the gap-filling limits of most anaerobics and it is better to use an epoxy resin (Araldite or similar) but these are high-strength adhesives and will probably make future disassembly difficult or impossible. If you are using either of these adhesives to clear this problem make it the final step in the overhaul process and allow the adhesives to cure with the motor fully assembled and mounted with the shaft axis vertical — this gives the best chance for capillary forces to centre the bearing in its housing.

4.2 Commutator motors

Commutator motors have the same mechanical problems as induction motors but the brushes and the commutator may also need attention.

Carefully inspect the commutator and brushes. Withdraw each brush from its holder being careful to observe its orientation. Check that it is a reasonable length, that it moves freely in its holder and that the business end is properly formed to the shape of the commutator and free from chips or erosion. If the brushes are OK and the commutator is an even blue-black colour, without obvious signs of wear, refit the brushes being careful to maintain their original orientation. This is a classic case of 'if it ain't broke don't fix it'. If a motor passes this inspection the commutator and brushes are nicely bedded in and are best left well alone.

If the brushes are in poor condition they should be replaced and bedded in by wrapping a small strip of 400 grit silicon carbide 'wet or dry' paper round the commutator (grit side out!) and working it back and forth until the brush end takes up the curvature of the commutator. If the correct brushes are not available look for a larger brush intended for a motor of similar operating voltage — low-voltage motors need fairly soft high-conductivity brushes, high-voltage motors need a rather harder brush material and need not be quite so conductive. Brush material can be machined with carbide-tipped tools but it is difficult to hold the brush securely without cracking it. Unless a large amount of material needs to be removed, it is usually simpler to rub it against a sheet of 60 grit 'wet or dry' paper stretched over a flat surface.

If you are replacing brushes, be sure the commutator is in good condition. If it is at all worn it is worth reconditioning it because new brushes will both bed in quicker and last longer on a freshly machined commutator. Brand new spare motors will have bright copper commutator bars. After a few hours of use the area swept by the brushes settles down to an even blue-black colour. All commutator bars should have a similar appearance. One or more bars unusually worn or a different colour indicates a faulty armature. A commutator will normally outlast several sets of brushes so a badly worn commutator is an indication of a faulty motor, or one that has seen a great deal of use. Badly worn commutators can be skimmed to provide a new surface but, unless there is a good reason for rescuing that particular motor, it is better to look for a more promising sample.

If you decide to skim a commutator take great care to ensure that the new surface is truly concentric with the armature shaft. Most motors have 60° conical centres cut into each end of the shaft so that it is both possible and convenient to true the commutator with the armature mounted between centres.

Some lack this useful feature. In this case support the commutator end of the shaft in a fixed steady and grip the drive end with a collet chuck or reasonably accurate three-jaw self-centering chuck. The obstruction posed by the fixed steady will usually make it impossible to reach the commutator with a lathe tool so use an appropriately sized combination centre drill to cut a new centre in the exposed end of the shaft. The fixed steady can now be removed and the commutator end of the shaft supported from a hard centre mounted in the tailstock.

If you haven't got a fixed steady, bolt a lump of hardwood to the cross-slide. Lock the cross-slide so that it cannot accidentally be moved. With a drill mounted in the headstock chuck use the normal carriage traverse to drill a hole right through the hardwood. Use a sharp drill of the same nominal size as the shaft at the commutator end. Much depends on the drill, but in most cases the fibres of the wood close up slightly when the drill is removed. You are then left with a hole which is exactly on centre height and a very light interference fit on the shaft end. This hole replaces the fixed steady and the rest of the operation can proceed exactly as described in the previous paragraph.

Recommended tool geometry for general purpose machining of pure copper is 7° front clearance, 7° side clearance, 10° back rake and 25° side rake, HSS tool at surface speed of 100 FPM to 200 FPM. In practice, at the very light cuts typical of commutator skimming, the tool geometry is not at all critical and if you simply grind up a tool with plenty of top rake and front clearance (about 20° and 10°) it will cut fine. Two important points are the tip radius and the fact that the tool must be really sharp. Fine grind

the tool to the shape shown in Figure 4.1 and then stone a small radius − 1/64″ or less round the cutting corner. This is the only part of the tool that does any cutting so be sure that it is really sharp

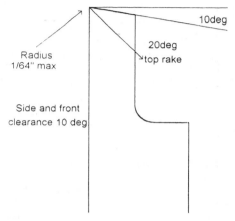

Fig. 4.1 *Tool geometry*

and that the 10° front clearance is maintained right up to the cutting edge − it is better to have too much rather than too little front clearance.

Machine at about 200 FPM, 0.005″ depth of cut and fine feed until the old worn surface is completely removed. Then take the final finishing cuts at not more than 0.001″ infeed. At these very light cuts cutting fluid is not needed as a coolant but a little light oil or soluble oil may make it easier to obtain a good finish.

Inspect the commutator carefully after machining. Small slivers of copper may bridge or intrude upon the intersegment separators and these must be carefully removed. Care is very important here as it is only too easy to accidentally scratch the freshly machined commutator surface.

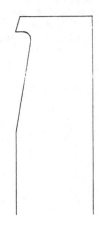

Fig. 4.2 *Hook tool*

It is slightly easier to guide a tool that is pulled rather than pushed so a broken hacksaw blade ground to the hook shape shown in Figure 4.2 is useful.

Motor manufacturers may or may not undercut the intersegment insulation. High-voltage motors with mica intersegment insulation are usually undercut. Low-voltage motors or motors with moulded plastic insulation are often left flush. Be guided by the construction in the unworn part of the commutator surface. Commutators that are intended to have undercut insulators usually have sufficient initial undercut to survive one or two skimming operations. If the commutator is a type that needs undercutting, and it is also badly worn, carry out the necessary deepening of the undercuts *before* skimming. Accidental scratches on the copper are almost inevitable when digging away at the undercuts but these are then automatically removed by the subsequent skimming process.

CHAPTER 5

Speed Control

5.1 General

Many items of workshop equipment need to be operated over a range of different speeds and variable speed motors are a very convenient means of providing this facility. However, not all motors are suitable for speed control and, even on suitable types, operation below the rated design speed necessarily sacrifices output power. Motors are essentially constant torque devices – for a given frame size the maximum torque that a motor can deliver is almost independent of speed. It sounds very attractive to have a 10:1 speed range but at the lower end of its speed range the motor can only deliver $\frac{1}{10}$ of its full-speed rated power. If a wide range electronic variable speed system is to be used to eliminate or reduce the need for mechanical speed changes then an appropriately larger motor is needed to provide sufficient power at lower speeds.

In this section it is impossible to avoid the use of some electronic jargon. In most cases the meaning should be evident from the context – in addition, a list of definitions can be found in Appendix 4.

5.2 Induction motors

In normal operation, the shaft speed of both single- and three-phase induction motors is controlled by the supply frequency and the number of poles in the stator (see Ch. 1, section 1.2).

The maximum speed cannot exceed the synchronous speed corresponding to the supply frequency. If an attempt is made to operate at reduced speed by reducing the supply voltage, the initial speed change will be quite small as the motor tries to stay close to its natural synchronous speed. With further reduction the motor speed falls to the speed at which it develops maximum torque – usually about two-thirds of the synchronous speed. With any further reduction the torque developed is insufficient to drive the load and the motor abruptly stalls.

The above comments apply to the sort of induction motors normally used to power workshop equipment. Some types of induction motors are manufactured for use in servo mechanisms or to drive variable speed overhead fans. These are special-purpose low-power machines that make heavy sacrifices in efficiency to obtain the convenience of voltage-controlled speed. The nearest common equivalent to these is the small shaded-pole motor. These can be voltage controlled over about a two to one speed range if they are operated with very light loads or just driving a small fan.

For motors suitable for the general run of workshop machinery the choice is limited to pole changing motors or operation from electronic variable-frequency, controlled-voltage inverters. These change the supply frequency to the value needed for the desired output shaft speed.

Pole changing motors are single- or three-phase motors provided with independent sets of windings that enable them to operate as normal induction motors at two or three fixed speeds. The speed change is made possible by distributing the windings round the stator so that each set of windings generates a different set of magnetic pole pairs — typically 2, 4 or 6 pole with corresponding full-load speeds of 2,850 rpm, 1,425 rpm and 950 rpm. They are rather larger and more expensive than their single-speed counterparts. Now that electronic variable-frequency speed control is becoming more affordable this is the more commonly chosen option.

These electronic controllers first rectify the single- or three-phase input power to generate a DC supply and then use digitally-controlled semiconductor switches to convert this to AC power at a frequency controlled by an internal oscillator. The control system also adjusts the output voltage to the value required by the motor for each output frequency and corresponding shaft speed.

Systems of this type are basically capable of driving single-phase or three-phase motors but are usually configured to drive standard 240V delta-connected three-phase machines as this avoids the complications of the special starting circuits needed by single-phase motors when operating from variable-frequency supplies. For motor powers of up to a few hp the electronic controllers operate from 240V single-phase supplies so, with three-phase output, this is also a convenient way of operating three-phase motors from single-phase supplies.

With both the frequency and output voltage under direct electronic control it is easy and inexpensive to add bells and whistles to this type of inverter. Typical features are:

    Automatic overload protection
    Torque limiting
    Acceleration limiting
    Preprogrammed speed profile
    Soft start option
    Remote speed control

Apart from overload protection (which is universally provided) and remote speed control these features are very limited value to the home user and should not be allowed to divert attention from the usually dominant constraints of price and physical size!

These controllers control the motor speed from the motor's normal 50Hz rated speed down to less than 100 rpm. Both the torque developed and the power dissipated within the motor are approximately constant over the whole of this speed range. If the motor is fan cooled it may be necessary to add a separately driven external fan to cool the motor when operating at the lower end of the speed range because the internal fan cannot shift enough air at low speeds.

Some controllers allow the motor to operate at speeds above the 50Hz rated speed — typically up to twice this value. However, both the motor power and output torque reduce when the set speed is increased, so that this higher speed range is only useful on comparatively light loads.

Remote speed control is a useful option. Few home workshops ever operate more than one machine at a time so it is perfectly practicable to use a single controller to supply a number of machines — a typical combination could be a lathe, a mill and a pillar drill. Simple on/off switching controls the power distribution but it is very useful to have the option of a remote speed control mounted on each machine.

For lathe and milling machine use, the electronic controllers are probably the most desirable speed control system of all — they are quiet, reliable, easy to install and cover a wide speed range with good speed regulation. The downside is that, in spite of more affordable pricing, they still remain the Rolls-Royce of speed control systems. The average home user will find it hard to justify the expense and will look long and hard at cheaper alternatives.

5.3 Commutator motors

5.3.1 General
It is much easier to control the speed of commutator motors. By suitable choice of armature and field operating conditions the shaft speed can be varied from a few rpm to many thousands of rpm, mostly from simple home-brew controllers assembled from low-cost components. An enormous variety of control systems are possible but to keep life simple, a few types are described here which, between them, should cover the majority of home workshop requirements.

Inevitably many of these circuits require the use of small electronic components soldered to printed boards. This technique may be unfamiliar to some readers but there is nothing inherently difficult about it and your success rate should be high. The important thing to remember is that the tiny components which control hundreds of watts of load power can

only do this if they are connected as intended. One wrong connection or a short circuit between two printed board tracks can divert enough of the load power to instantly destroy one or more of the control components. If you are lucky a wisp of smoke may indicate which one but, in many cases, more than one component is damaged and there are no visible signs of failure. This turns what should have been a straightforward assembly and test operation into a lengthy and frustrating detective exercise.

The moral of this is to check, check and check again *before* applying power to the set up. Some of the set ups use stripboard with the components soldered to closely spaced copper tracks. Inspect the space between the tracks very carefully — it is easy to accidentally bridge this space with a tiny sliver of solder. Also check that all the required breaks in the tracks are present and in the correct position.

5.3.2 General-purpose triac controller
This is a simple triac controller of the same general type that is fitted to portable electric drills, food processors and similar domestic variable-speed devices. Figure 5.1 shows the circuit arrangement and component list. Figures 5.2 and 5.3 show the mechanical layout when assembled on standard copper track strip board.

Q2 is the main power control triac. This is a semiconductor device that is normally open circuit and prevents any power reaching the motor terminals. If triggered 'on' by a short pulse from Q1 it remains on for the rest of that half cycle of the supply frequency. It automatically switches 'off' as the input supply voltage passes through zero and does not switch on again until another pulse is received from Q1.

Triacs are bidirectional devices — they

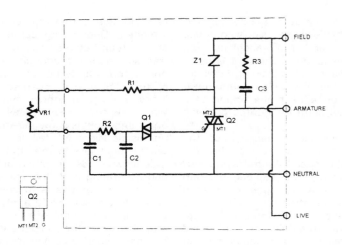

R1 4.7 kohms 0.6W	Maplin M4K7	C1 0.1 μF 250V	Maplin BX76H
R2 22 kohms 0.6W	Maplin M22K0	C2 0.033 μ 250V	Maplin BX73Q
R3 100 ohms 0.6W	Maplin M100	C3 0.01 μ 250VAC	Maplin JR31J
VR1 220 kohms 0.4W	Maplin FW06G	Z1 250L 250VAC	Maplin HW13P
Q1 DB3 Diac	Maplin QL08J	Q2 C246M 600V/16A	Maplin UR31J

Hardware

Stripboard 2939	Maplin JP47B	Spade connectors (M)	Maplin AS33L
Heatsinks	Maplin FG55K	Spade connectors (F)	Maplin AS30H
		Connector covers	Maplin AS31J

Fig. 5.1 *GP triac controller*

can be switched on by a positive or a negative pulse at the gate electrode and, when on, can pass current during both the positive and negative half cycles of the supply frequency.

Q1 is a bidirectional trigger diode usually called a diac. It is a semiconductor device that is normally open circuit but suddenly breaks down and turns itself on if the voltage across it reaches 32V in either direction. In this circuit it is used to discharge C2 into the gate of Q2 and thus switch Q2 on whenever the voltage on C2 reaches 32V.

VR1, R1, C1, R2 and C2 are the components that control the time in each half cycle of the supply frequency that the voltage on C2 reaches the level needed to switch on Q1 and Q2. If VR1, the speed control, is set near maximum resistance Q1 and Q2 do not switch on until almost the end of each half cycle. They are off most of the time and the average voltage reaching the motor is very small. If VR1 is set to zero, Q1 and Q2 switch on almost at the very beginning of each half cycle and the full mains voltage reaches the motor.

C1 is the main timing capacitor in this network — the value chosen allows VR1 to control all the way from zero output to full power. Higher or lower (up to about

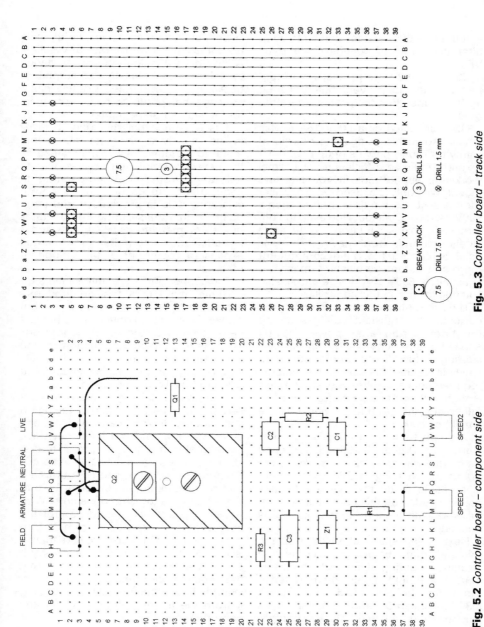

Fig. 5.2 Controller board – component side

Fig. 5.3 Controller board – track side

63

+ 100% or − 50%) values can be used to alter the range of speeds covered by VR1.

C3, R3 and Z1 are protective components added to prevent fast transient voltages upsetting the proper operation of Q2.

Q2 is specified as a 600V 16A device which provides an adequate margin of safety for motor powers of up to about 1 hp and should be sufficient for all normal workshop applications. If you are sure that it will only be used with smaller motors such as spin drier motors, then a 600V 8A device will do. However, the difference in cost is not large and the extra safety margin is worth having. The triac used to control the main commutator motors in automatic washing machines is usually a 600V 12A device and is also suitable for use in this controller.

The triac only has to handle large peak currents and transient voltages during the brief period after power is first applied to a motor while the armature is starting to turn. Once normal operating speed is reached the current and voltage requirements are quite modest. The triac is then working at a small fraction of its full-power ratings so that it only needs a small heatsink.

The construction is quite straightforward − first drill the additional holes and then cut away the copper tracks at the points shown in Figure 5.3. A special tool is available to do this but for the few breaks needed on this board a little careful work with a sharp knife is all that is needed.

Next fit the spade connectors to the two ends of the board. These come with straight mounting pins which should be bent over at right angles and mounted in the 1.5mm holes marked XX in Figure 5.3. The insertion and removal forces can be rather high with this type of high-current connector so it helps to lubricate the contact faces with a trace of Vaseline. A 25W or 40W soldering iron is needed to solder these pins to the stripboard tracks. If you are limited to the smaller 15W size then *gently* preheat the connector tab with a butane lighter.

It is essential that the triac is firmly bolted directly to the heatsink so that metal to metal contact is maintained as the heatsink temperature cycles over its full range from room temperature to about 100°C. To ensure this, a 7.5mm clearance hole is provided in the stripboard so that the securing bolt operates directly on metal-to metal surfaces and is not affected by the rather large thermal expansion coefficient of the stripboard material.

The copper tracks are only suitable for currents of up to a few hundred milliamps so the triac connections are not routed through the copper tracks but soldered directly to the spade connectors. There is no suitable termination for the gate connection of the triac so a flexible wire is soldered to the triac leadout and taken to hole B9 on the stripboard.

With the remaining components inserted and soldered in, inspect the board very carefully looking particularly for any point where a soldered joint may have accidentally bridged adjacent tracks. Once you are sure that all is well, to test it, connect the VR1 speed control to the end connectors and a 60W light bulb to the armature and field connectors. With power applied RV1 should control the lamp brightness smoothly from zero to full brightness. Once you are sure that all is well on a lamp load it is safe to try it out on a suitable motor. Don't forget that the whole board (including the heatsink!) is live at mains potential − unplug it from the mains before touching any part of it.

This sort of controller is fine for the smaller series-wound universal motors driving a fairly constant load — it is the type used to drive the power fretsaw described in Chapter 9, section 9.7. However, it simply controls the average power input to the motor and takes little account of motor speed. At maximum power setting an unloaded series-wound motor can still overspeed.

5.3.3 Series-shunt controller

A big improvement in speed regulation can be obtained by rectifying the output of the GP triac controller and using it to supply either a permanent-magnet motor or a wound-field motor provided with a separate field supply.

240V permanent-magnet or shunt-wound machines are not too common. However, series-wound machines are easily converted to the equivalent shunt-wound machine by supplying the field current from a separate rectified low-voltage supply. The ubiquitous automatic washing machine motor is normally series connected, but it is quite easy to convert by this method.

The arrangement is shown in Figure 5.4. The low-voltage transformer is chosen to supply about the same field current as the load current of a fully loaded motor. The armature current is still routed through the series-field winding as this has the twin benefits of both reducing the peak starting current and also increasing the torque. On very heavy mechanical loads, once the armature current exceeds the shunt-field current, the extra series-field current results in more field ampére/turns and this increases the maximum torque.

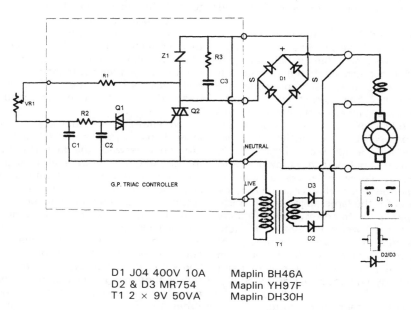

D1 J04 400V 10A	Maplin BH46A
D2 & D3 MR754	Maplin YH97F
T1 2 × 9V 50VA	Maplin DH30H

Fig. 5.4 *Series-shunt controller*

Be very careful to ensure that the motor is connected as in Figure 5.4 which routes the low-voltage shunt-field current through the field windings in the *same* direction as the armature current. If either of the rectifier connections are reversed the armature current will oppose the shunt field — the motor will fail to start and probably take enough current to blow your fuse protection! Of course, it is OK to connect the field winding either way round to suit your choice of direction of rotation.

This controller plus one of the larger ex-automatic washing machine motors is a useful variable-speed set up for driving the average small workshop lathe, milling machine or drill press. If replacing a 1,425 rpm induction motor an additional speed reduction of about 1:3 to 1:4 will be needed to enable the new motor to operate in the middle of its useful speed range.

A gear reduction stage is likely to be too noisy and the preferred choice is a poly-vee belt or a timing belt drive. A 0.2"/5 mm tooth pitch timing belt will easily handle the power provided the motor pulley has more than 18 teeth. Teeth are an unnecessary luxury on the larger driven pulley as, at this diameter, friction will provide sufficient drive torque with a flat-faced plain pulley. Since neither pulley face is crowned both pulleys should carry flanges to keep the belt centred. It is perhaps natural to think of metal for the large pulley but at these belt speeds the forces involved are quite small and plywood pulleys have more than sufficient strength. If the thought of wood on your machine offends you, then spray the pulley with aluminium paint and nobody will know the difference!

5.3.4 Low-voltage triac controller
This very simple controller is useful for controlling small motors operating from low-voltage transformers. The principal disadvantage is that it is only suitable for use with permanent-magnet machines, or wound-field machines with a separate fixed field supply. Series-wound universal machines do not generate enough back EMF for it to work properly.

The arrangement is shown in Figure 5.5. This time the triac is triggered 'on' by the voltage at the slider of RV1. Over several cycles of the supply frequency the motor accelerates until it reaches the speed when the back EMF that it generates exceeds the peak voltage at the slider of VR1. This reverse biasses all four diodes of D1 which prevents the trigger electrode from switching the triac on again at the next positive or negative half cycle. The triac will now remain off, until the speed and the corresponding back EMF drops enough to allow the triac to switch on again.

At low-set speeds and light loads the triac will remain off for a number of cycles of the supply frequency. At higher speeds and loads the triac switches on at every half cycle but the length of each 'on' period depends on the motor speed and load.

The length of time that the triac is on in each half cycle is controlled by the difference between the back EMF of the motor and the set voltage at the slider of the speed control potentiometer. This is a form of closed-loop feedback and results in good speed regulation.

In this circuit, the downside is that it is not possible to switch the triac on for less than a quarter of one cycle of the supply frequency. On very light loads, at low-set speeds, a quarter cycle may contain enough energy to cause the motor to accelerate well beyond the set speed. When this happens, the triac controller switches off and waits for the speed to drop back to the set level before it provides another pulse, so that the motor hunts between the set speed level and a higher level. This only occurs at low-speed settings and disappears as soon as either the set speed is increased or the mechanical load on the motor is heavy enough to limit the speed increase from a single impulse.

In practice this is only really bothersome on a lightly loaded motor at low speeds. If the motor is delivering significant power to a load, the effect is either absent or the impulses are so closely spaced that the speed of the motor is almost constant.

This circuit relies on the fact that most triacs need very little current to trigger them on — much less than the manufacturer's guaranteed limit. I have checked this circuit out with a wide range of triac types from several manufacturers with current ratings of up to 16A. Most worked fine so you are unlikely to have problems. If you should be unlucky enough to get a triac that needs a excessive trigger

current the motor will probably only run over the top quarter rotation of RV1 and may not reach full speed.

The values in Figure 5.5 are chosen to suit the input voltage range 9V to 30V. Below 9V the voltage needed to switch the triac on is too large a fraction of the supply voltage. Above 30V too much power is dissipated in VR1.

The triac type should be chosen to suit the motor — a 400V 8A device should be sufficient for most applications.

When switched on there is a voltage drop of a bit less than 2V across the triac. For anything over 1A continuous full-load current the triac will need to be bolted to a heatsink to get rid of the heat that this generates. For load currents of up to 6A the small heatsink used in the general-purpose controller mentioned in section 5.3.2 is adequate. For higher currents a larger heatsink can be used but it is usually more convenient to bolt the triac directly to the metal case of the controller so that the whole case acts as

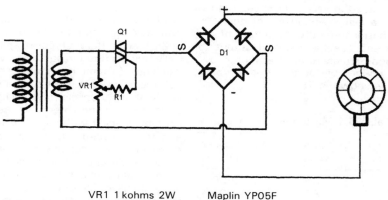

VR1 1 kohms 2W   Maplin YP05F
R1 100 ohms 0.6W   Maplin M100
D1 PW01 100V 6A   Maplin WO58N
Q1 CF225D 400V 8A   Maplin UR36P

Fig. 5.5 *Low-voltage triac controller*

a heatsink. The mounting faces of most triacs are 'live' so they must be insulated from the metal case. Insulating kits are available (Maplin Kit P Plas, WR23A) but, unless you already have the triac, it is more convenient to use the CF225/246 series of triacs which are already fully insulated and can be bolted directly to a case without additional hardware.

5.3.5 Variac speed controller

For the experimenter who is determined not to get involved in electronics this is an extremely simple, but more expensive, speed control system. A Variac is a variable transformer that can operate from the 240V AC supply and deliver an output that is continuously variable from zero volts up to 240V. On some types there is an alternative tapping point that then allows the output to be varied from 0V to 270V. Maplin DM96E or DM97F are suitable for this use.

They can be used to directly control a series-wound universal motor without any additional equipment. However, this perpetuates the very poor speed regulation of a series-wound motor and it is much better to rectify the output of the Variac and use the rectified output to control a permanent-magnet motor.

The no-load performance on a typical motor is shown in Table 5.1

Table 5.1 Motor performance

AC input volts	DC to motor volts	Shaft speed rpm
50	56	1,600
75	87	2,900
100	119	4,100
150	187	6,600
200	251	8,900
240	303	10,700

At light loads, when the motor is connected to the rectifier, the DC output voltage rises and is higher than the AC input voltage. This happens because on light loads, the motor tries to accelerate to the speed corresponding to the peak value of the rectified DC. It doesn't quite achieve this but it reaches a speed high enough for the back EMF to exceed the normal average value of the rectified output. It is this that is responsible for the higher measured voltage at the motor terminals. The motor behaves rather like a capacitor – it only draws current for short periods near the peak value of the input voltage and this causes the average value of the voltage at the motor terminals to lie somewhere between the RMS and the peak (see Appendix 4) value of the input voltage.

Both the speed and the back EMF drop as the load on the motor increases and at full load, the motor will be drawing current for most of the time. The rectified voltage and speed can be expected to be about two-thirds of the no-load figures.

5.3.6 Commercial speed controllers

You may be tempted to try to adapt one of the many different types of speed controller boards fitted to automatic washing machines. In some cases this is possible but it is an exercise that can be difficult and frustrating. Unless you are a hardened electronic experimenter it is better to content yourself with recycling the triac and the transient suppressor. Both of these items are suitable for use in the controller described in section 5.3.2 .

The main problem is that the boards are optimised for operation at two or three widely separated fixed speeds and not intended for variable-speed operation. In addition, there are many different types of board and a fix suitable for one may be useless on another.

The following pointers may be helpful.

(a) There will be two heavy current spade tags usually marked 'N' and 'F'.

N goes to 240V AC neutral. Wiring loom colour code — grey/blue.

F goes to one motor field connection. Wiring loom colour code — red/blue. The remaining field connection links to one armature connection.

The remaining motor armature connection goes straight to 240V live.

(b) There will be a multiway connector — usually 16 pin. One or more of pins 14, 15, 16 will go to a high power resistor. This is the 240V live connection — the cable loom wire is usually colour coded white/blue.

(c) At the opposite end of the connector, pins 1 & 2 are the tacho generator connections — both grey wires in the cable loom.

(d) Three or four cable loom wires go to the middle pins of the connector. These are the control pins and each of them goes directly to separate high value resistors. Usual colour codes — green/black, brown red,black and grey/black. Green/black to the almost adjacent 240V live pin switches the motor on. 240V live to one of the other pins selects the operating speed mode.

(e) Because triacs prefer negative going trigger, the low-voltage rail is negative with the common positive connected to 240V neutral.

Not all modules conform to these connections but many of the types that use the popular TDA1070 control IC are generally similar.

The high-speed spin mode is the useful one but it includes a soft start feature which takes many seconds to bring the motor up to speed. With the TDA1070 this is controlled by a 47 μF or 100 μF capacitor connected to pin 13 — this

should be reduced to 1 μF.

The timing capacitor (usually 0.1 μF) which determines motor speed connects to pin 10 via a small signal diode. The other end connects to the slow-start capacitor. Some variation of the controlled speed can be made by varying timing resistors associated with this part of the circuit, but wide-range speed control is better carried out by switching in different values of timing capacitor — 0.068 μF to 1.0 μF is the useful range.

Most of the examples in this book have been thoroughly tested and, unless you have been very unlucky with your particular sample of a motor type, should work as described. However, this does not apply to this section. If you happen to find the right type of control board this set up can work very well indeed. But equally you may well waste a lot of time on an unsuitable module and finish up with no useful result!

5.3.7 DC controllers

The controllers so far described all rely on the use of an AC supply to switch triacs or operate transformers. Although special switching circuits can be used to enable SCRs and triacs to operate from DC supplies, the operating conditions need to be tightly controlled and are not very suitable for general purpose home-brew controllers.

A better type of device for battery and DC supplies is the power MOSFET. This is rather similar to the more familiar bipolar power transistor but has the advantage it can be switched on or off by a small voltage applied to the gate electrode. The gate electrode requires little or no current to operate it, so simple low-power control circuits can be used, even for power MOSFETs that are handling many hundreds of watts. Bipolar power transistors can be used in similar applications but require large currents (typically ¹/₁₀ of the output current) to switch them on and this complicates the drive arrangements.

Figure 5.6 shows the circuit arrangement for a general purpose MOSFET speed controller for operation from a 12V battery or 9V to 16V DC supplies. IC1 is the very popular 555 timer set to oscillate at about 10kHz. VR1 controls the duty cycle — the fraction of the time that the output on pin 3 is high. When the slider of VR1 is near R2, pin 3 is near supply voltage for about 98% of the time.

With the slider near R1, pin 3 is high for less than 2% of the time.

The MOSFET Q1 is only on when its gate voltage is high. It is nearly a perfect switch — the 'on' resistance is only 0.03 ohms, so the average value of the output voltage is almost exactly the supply voltage times the duty cycle. The switching frequency is so high that the inductance of the motor armature smooths this to almost pure DC so that the controller acts as a duty cycle controlled, variable-voltage source. This means that, if it is used to control a permanent-magnet or

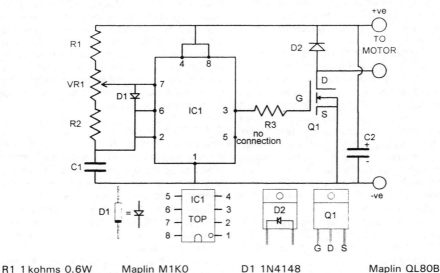

R1 1 kohms 0.6W	Maplin M1K0	D1 1N4148	Maplin QL80B
R2 1 kohms 0.6W	Maplin M1K0	D2 MBR745	Maplin GX31J
R3 150 ohms 0.6W	Maplin M150	IC1 NE555N	Maplin QH66W
VR 100 kohms 0.4W	Maplin FW05F	Q1 BUZ11	Maplin UJ33L
C1 0.01 400V	Maplin BX70M		
C2 1000 μF 50V	Maplin JL57M		

Hardware

Stripboard 2939	Maplin JP47B	Spade connectors (M)	Maplin AS33L
Heatsink	Maplin FG55K	Spade connectors (F)	Maplin AS30H
		Connector covers	Maplin AS31J

Fig. 5.6 *MOSFET controller*

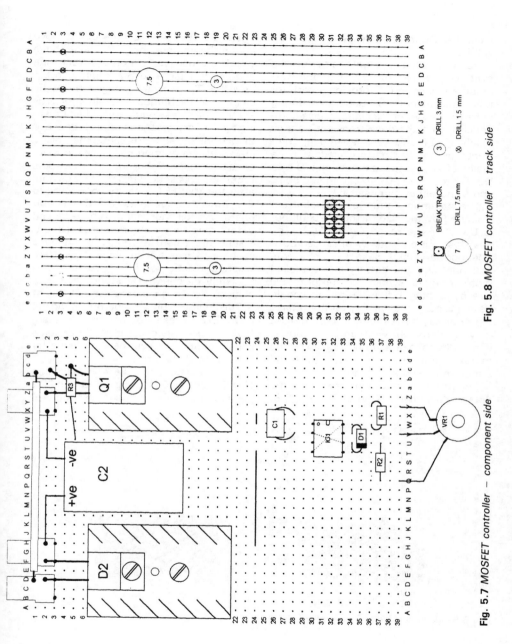

Fig. 5.8 MOSFET controller – track side

Fig. 5.7 MOSFET controller – component side

71

fixed-field DC motor, the speed regulation will be good.

R3 is included to slightly slow up the switching speed of Q1. D2 and C1, mounted close to Q1, provide a path for the transient currents generated during parts of the switching cycle. These components limit the peak transient voltage reaching Q1 and keep switching transients out of the battery leads. To be really effective the connections Q1 to D2, D2 to C2 and C2 to Q1 should be as short and direct as possible.

The maximum continuous-load current is limited by the size of the heatsink but is OK over the range 0A to 10A. The peak-load current should not exceed 30A. The normal worst case peak currents are: fully charged battery into a stalled motor, or rapid reversal of a motor that is already running at full speed. This should be ample current capability for the smaller motor projects but the peak-current limit may be a problem with larger motors – if in doubt check the current taken by the motor directly from the supply with the armature locked stationary.

Construction is straightforward – the details are shown in Figures 5.7 and 5.8.

There are four links on the board. Two are on the component side – just above C1. The other two are on the track side and link IC1 pin 4 to 8 and pin 2 to 6.

MOSFETs can be damaged by static electricity discharges – typical hazards are the tiny sparks generated when brushing against dry synthetic fabrics. Because of this, new items are sometimes delivered with conducting black plastic foam or silver paper shorting all the pins together. This should be left in place until after the MOSFET is safely connected to the other components on the circuit board. Tin-foil shorts should be removed before applying power to the circuit. Plastic foam can be removed but does no harm if it is left in place. If you are wearing plastic- or rubber-soled shoes you may be carrying a static electricity charge. Touch an earthed metal object to discharge yourself before handling MOSFETs. These power MOSFETs are fairly rugged devices and static damage is uncommon – nevertheless it is better to be safe than sorry!

With modifications the controller can also be used on 24V supplies. Q1 and D2 need to be higher voltage versions –

Fig. 5.9 *Triac and MOSFET controllers*

IRF540 and MBR1090 are suitable items. The 555 timer is limited to 16V max. The easiest way round this is to feed pins 4 and 8 and R1 from the output of a 7812 voltage regulator. This accepts 24V from the supply and delivers 12V to the 555 circuits. The power consumption is very small so the 7812 does not need a heatsink.

The standard controller will also work below 9V but the MOSFET may not be fully switched on when delivering large currents. Set VR1 to maximum output and measure the DC voltage between S an D of the MOSFET. This should be no more than 1½V when delivering your largest load current.

CHAPTER 6

Mobile Power

6.1 General

Most of this book is concerned with the use of motors as power sources in the home workshop. A slightly different field which is of interest to many model engineers is the use of battery-powered motors for 3½" and 5" gauge electric locos, children's buggies or electric car projects. Each of these applications can be fairly major undertakings so detailed description is beyond the scope of this book — my own experience in this field involved a home-brew electric wheelchair which was a reasonably successful project but it occupied a full year of my spare time! However there is much common ground in the power train arrangements for the larger locos, the buggies and the electric cars and a few comments on the motive power aspects may be helpful.

6.2 Motors

Generally, the motor should be the first item to be chosen as the motor power rating is dictated by the performance required from the vehicle. The chosen motor, in conjunction with the vehicle range or operating time, then defines the drive train and battery capacity requirements.

This is, of course, in an ideal world with a free choice of all components. In practice, the starting point is more likely to be "here's a likely looking motor — how do we make best use of it!".

For home workshop projects the range of suitable motor power ratings is quite small. If the loco, buggy or car is to be capable of carrying at least a one-person payload over a reasonable range of terrain then ¼ hp is about the minimum practicable rating. If it's a buggy for a small child or a loco running on a level track then ⅛ hp is possible but marginal.

In the upward direction ½ hp is ample for almost all buggy and loco requirements but, without other limits, a little electric car could happily absorb many hp. However, high-power motors need correspondingly large batteries to supply them and the upper power limit is then severely constrained by the cost, weight and size of the necessary battery pack.

Batteries as the prime power source dictate the choice of a commutator motor both for its compatibility with a DC supply and its suitability for wide-range speed control. In principle, a motor of the series, shunt, compound wound or permanent magnet type could be used but a permanent-magnet motor is the preferred choice as it is slightly smaller, lighter and more efficient than its wound field counterpart. However, while small, low-voltage permanent-magnet commutator motors are plentiful, motors in this higher power range are much less common and you will be lucky to find a suitable item.

A source that is, at first sight, attractive is the large permanent-magnet motors used to directly drive the 10" reels of 1" magnetic tape in some of the older mainframe computer tape decks. These motors were chosen for their high torque at low speeds which provides the high acceleration needed for the rapid start/stop motion of the tape reels. Typically, they operate from a transistor controlled 24V supply but at this voltage their maximum speed is often only 500 rpm. This is ample for their intended computer use because the reels never continue to rotate in either direction long enough for the motors to approach their supply voltage limited speed, but it severely limits the output power – at this low speed an impressively large motor is often only capable of 1/10 hp or less!

Of course, the performance can be recovered by increasing the supply voltage and these motors will comfortably deliver ½ hp at 3,000 rpm, but this needs at least a 100V supply which is a wildly inconvenient level for small battery-powered vehicles.

Another possible source is the wide variety of ex-aircraft and surplus military machines. These are mostly 24V machines which is fine for a vehicle that can comfortably accommodate a pair of 12V batteries, but not very suitable for a loco or a buggy where even a single 12V battery occupies a large fraction of the available space.

Some of the ex-aircraft machines are sufficiently small and light but are rated to deliver more power than needed for your project. In this case it is quite accept-able to operate from a reduced supply voltage. After allowing for the slightly reduced efficiency, a 24V motor operated from a 12V supply will deliver about 40% of its rated power at a little less than half its rated full-load speed. This assumes that the motor is operated with the same field strength and at the same full-load current.

If it is a permanent-magnet or series-wound field motor the field strength will still be correct and no change will be needed to the motor or its connections.

If it is a shunt- or compound-wound motor, the two shunt field coils which are normally connected in series should be reconnected in parallel so that the full supply voltage reaches each field coil. The same comment applies to the shunt-wound coils of a compound-wound motor but the series-wound coils will already be correctly connected and should not be disturbed.

The 40% power estimate is very conservative because aircraft and military machines are normally rated to operate at very high ambient temperatures – $55°C - 130°F$ or more. This is far higher than will be encountered in home-brew projects so the normal rated full-load current can be significantly exceeded without fear of the motor overheating. In many cases the duration of a sustained overload will be limited by a fading battery rather than an overheated motor.

Another possibility is the use of a car dynamo as a motor as described in Chapter 3, section 3.4. Although these have long been replaced by alternators, dynamos were originally fitted to a very large number of vehicles and usable survivors are not too difficult to find. The big problem here is the field supply – if a 12V dynamo, used as a motor, is to deliver anything more than a small fraction of its capability, the armature needs to run at 24V or more while the field supply must be maintained at 12V. The simple solution is to feed the field from the 24V supply via a dropping resistor although this doubles the power absorbed by the field circuit. A better solution is to feed the 12V field from the 24V main supply via a 2:1 step-down, duty-cycle regulator of the type shown in Figure 5.6. These operate at well over 80% efficiency so very little power is wasted in the 24V to 12V conversion.

At the time of writing there are still a few motors remaining from the ill-fated Sinclair C5 electric vehicle on the surplus and second-hand market. These are ideal for small vehicle projects but unless some entrepreneur decides to make some more, they will soon be unobtainable.

The comments so far have mainly referred to the use of a single-drive motor. However, if the available motors are on the small side there is a lot to be said for multiple-motor drive. Twin-motor drive, driving separate wheels, is popular on wheelchairs as differential control of the two motors results in a very simple form of power steering. It is also a convenient method of sidestepping the requirement for mechanical differentials on conventional two- or four-wheel drive vehicles but it carries with it the penalty of two or four separate drive trains to match the motor speed to the driven wheel speed. If more power is the only requirement it is much simpler to couple two or more motors to the single-drive belt providing the first-stage speed reduction.

The motors can be connected in series or in parallel. At constant field strength, if the motors drive independent wheels, series connection results in equal torque output, parallel connection results in equal wheel speed. If the motor shafts are closely coupled together, either directly or through the road, either connection

can be used and the load will be shared fairly equally between the two motors.

6.3 Batteries

Batteries are the Achilles heel of mobile electric projects. It is rarely a problem to carry enough fuel to achieve the desired range/endurance in a steam or internal combustion-engined project. This is not the case with electric-powered vehicle because the cost and weight of the battery makes it almost always necessary to accept severe limitations on range/endurance. If aiming for extended performance it is quite easy to be faced with a design in which the majority of the weight and cost of the vehicle is taken up by the battery — if this sounds extreme check the weight of the battery in your friendly neighbourhood electric milk float!

Currently there are plenty of 'gee whiz' articles on the wonder batteries of the future but the rate of progress on commercially viable items is disappointingly slow. The present position is that the only systems practicable for small projects are the well-established lead—acid and nickel—cadmium technologies. Nickel hydride (NiMH), and lithium ion rechargeables are becoming established in the portable computer and camcorder field but are only competitive at low and medium discharge rates. At the high discharge rates typical of vehicle projects, lead—acid and NiCads are still the best bet.

The small sealed NiCads (e.g. D and F sizes) are very convenient to use and, with proper precautions, tolerant of very high charge and discharge rates. But unless a very small vehicle is contemplated the maximum available capacity (less than 10Ah) is inadequate. Industrial unsealed NiCad cells are available in a wide range of capacities and are often used in electric fork-lift trucks and similar applications. While these are excellent batteries and stand up well to the arduous operating regime of repeated charge/deep discharge cycles they are too expensive for general use.

Special varieties of lead−acid batteries are made for traction use. These use a plate construction which is more resistant to damage from repeated charge/deep discharge cycles. While they are a little more affordable than NiCad traction batteries, for equal capacity, they are still substantially larger and more expensive than the readily available car starter batteries and general-purpose sealed lead−acid types. This means that unless you have a very deep pocket, or are lucky enough to find a small second-hand traction battery, you are limited to car batteries for the larger projects and sealed lead−acid for smaller items.

Car batteries are well suited for brief periods of rapid discharge but do not stand up well to repeated deep discharge cycles. The manufacturer's life guarantees are usually invalid if the battery is used other than as a starter/lighting battery in a normal road vehicle. To get the longest possible life take the following points into account:

(a) Use the largest battery that the project and/or your pocket can accommodate.

(b) If space constraints necessitate the use of two or more battery compartments, use low-voltage sections series connected e.g. use two 6V batteries in series not two 12V batteries used independently or in parallel.

(c) No matter how great the temptation don't over-discharge the battery. This means don't over-discharge any individual cell in the battery − it only takes one dud cell to ruin a battery!

(d) Don't leave the battery in a discharged state any longer than necessary. If possible, recharge immediately after use.

(e) If possible use constant voltage charger set to 14.4V. If an ordinary DIY charger is used, check the battery voltage near the end of charge and aim to stop the charge before it reaches 15V. In an open battery it is possible to replace the water lost by overcharging so occasional over-charging for limited periods is quite acceptable. However, if the battery is frequently overcharged for extended periods the lead matrix in the positive plates will start to swell and distort and this results in early failure.

(f) Don't allow the batteries to sulphate up when out of use. Sealed cells need six-monthly recharges. Open car batteries self-discharge more rapidly and need top-up charging at least once every two months.

Occasionally you may come across discarded traction cells or aircraft starter batteries. Lead−acid cells are usually beyond recovery but open (i.e. unsealed) NiCad cells in rectangular steel or plastic cases can sometimes be rescued. The basic electrode structure of these cells is extraordinarily long-lived and often it is only the electrolyte that is at fault. I have a 12V stack of 20Ah cells that are well over twenty years old and are still happily powering the starter motor on my tractor lawnmower.

The correct electrolyte is potassium hydroxide in distilled water SG 1.305 (approximately 31% by weight). Don't forget that this is very caustic − no skin contact, always wear rubber gloves and eye protection when handling electrolyte. There is a strong heat build-up when potassium hydroxide is added to water, so if you make up your own electrolyte, add the potassium hydroxide slowly to the distilled water and give it time to dissolve. **NEVER** pour water on to neat potassium hydroxide — it will boil and spit electrolyte in all directions!

Unlike lead—acid cells, the electrolyte in these cells remains chemically unchanged throughout the whole charge/discharge cycle and its primary function is to act as an extremely low-resistance conducting liquid. Unfortunately over a very long period (years) the electrolyte slowly reacts with carbon dioxide from the atmosphere and partially converts to potassium carbonate. This eventually raises the internal resistance of the cells to the extent that they can no longer meet the fast discharge requirements and are scrapped.

The obvious solution — to change the electrolyte — is not as easy as it sounds. Most of the electrolyte is absorbed into the pores of the porous plate material and less than a quarter can be poured away, the remainder staying in the cell. If fresh electrolyte is added it does not readily mix with the stale electrolyte retained in the plates until the cell is brought up to a full state of charge and is gassing freely. Four or five electrolyte change and charge cycles are needed to get rid of most of the old electrolyte. At each change the charger should be switched off when the cell is gassing freely and the electrolyte quickly poured off while the pores in the electrodes are still full of gas. Refill with fresh electrolyte and repeat the cycle. The gas is, of course, an inflammable hydrogen/oxygen mixture so carry this out in a well-ventilated area and ensure that there is no chance of accidental ignition — *be sure that you switch off the charger before breaking any connections*!

This is quite a major operation with a set of traction cells but it is not too difficult with the smaller cells used in aircraft starter batteries. These are usually a stack of small individual cells that can be handled separately and each change of electrolyte in each cell needs less than 50 ml of fluid.

CHAPTER 7

Battery Power Supplies

7.1 General

Many of the smaller commutator motors are operated from batteries. Primary batteries are a very convenient, but spectacularly expensive, source of power ranging from about £200 per kilowatt-hour for the larger sizes to £5000 per kilowatt-hour for the small, silver-oxide wristwatch cells! Secondary (i.e. rechargeable) batteries are more economic. Based on a life of 500 charge/discharge cycles the cost drops to less than £10 per kilowatt-hour. This cost is based on the purchase price divided by the number of charge/discharge cycles — at a few pence per kilowatt-hour the cost of the recharging power from domestic supplies is negligible.

In the following notes the terms 'high current' and 'low current' occur frequently. These terms are related to the capabilities of the cell under discussion. As a rough guide, a high current will discharge a cell in less than one hour while a low current can be maintained for at least one hundred hours. No prizes for guessing that a ten-hour life will result from a moderate load.

The capacity of cells is often quoted in ampère/hours (Ah) or milliampère/hours (mAh) — the product of the current delivered by the cell and the time it takes to discharge it at that current.

7.2 Primary cells

The main types of primary cells are listed in Table 7.1

Table 7.1 Primary cells

Generic name	Popular name	Initial voltage	End life voltage	Shelf life	Cost
Leclanche	Zinc carbon	1.5	0.9	1–2Y	low
Zinc chloride	High power	1.5	0.9	1–2Y	low
Alkaline	Alkaline	1.5	1.0	3–5Y	medium
Mercury	Mercury	1.35	1.1	3–5Y	high
Silver oxide	Watch	1.55	1.1	3–5Y	high
Lithium	Lithium	3.0	2.0	>10Y	high

7.2.1 Zinc carbon and alkaline cells

There are three types in this category and, between them, they represent the vast majority of the primary cells in common use. They are all initially 1.5V cells and this voltage drops steadily in use until they reach the end of their useful life at about 0.9V. They are manufactured in the same range of external shapes and sizes and, in many applications, any of the three types can be used.

The widely available basic Leclanche/zinc carbon type is suitable mainly for low-current loads or intermittent use at higher currents. Typical uses are as torch batteries or for powering small transistor radios.

Operation at very low currents is limited by the comparatively short shelf-life and in critical applications it is advisable to replace the battery at least annually, even if the current drain is negligible. Storage temperature has a marked effect and a temperature increase of less than 10°C will halve the shelf-life.

These cells are only suitable for operation at high currents if used on an intermittent basis. At high currents, gas is generated at the electrodes faster than it can be absorbed by the chemicals in the cell. This gradually increases the internal resistance of the cell and the output voltage falls with an apparent loss of capacity. In intermittent use most of this lost capacity is restored after a period of rest which enables the excess gas to be absorbed.

The zinc chloride/high power cell is a variant of the zinc carbon cell with a lower internal resistance and a higher rate of absorption of excess gas. It is only a little more expensive than the standard zinc carbon cell but is more suitable for motor loads because of its improved high current performance. At low currents it is similar to the standard cell.

Both types of cell use an acidic electrolyte contained in a cylindrical zinc negative electrode. Most modern cells surround this with an outer protective plastic or metal cover. When a cell is completely discharged, the zinc electrode can no longer resist the attack of the electrolyte which eats its way through the zinc and tends to leak past the cover. The cover is usually successful in containing this for a month or so but after that leaking cells can do a lot of damage.

These cells are not hermetically sealed and even in normal use a very small amount of electrolyte can find its way onto the adjacent battery contacts and wiring. This produces bright green corrosion products on unprotected copper and many copper alloys. Commercial battery holders mostly use nickel-plated steel which is fine providing the plating thickness is adequate. Stainless steel and nickel silver are good materials for home-brew contacts. Stainless steel is very good from the corrosion resistance point of view but the contact pressure must be high enough to break through the surface oxide film that is always present. A good source of nickel silver is the contact springs of old relays.

Don't use brass or phosphor bronze as the green corrosion products will soon generate high contact resistance. Occasionally you may come across a piece of equipment that, through ignorance or expediency, breaks this rule. Cleaning the contacts will only give short-term relief – the best solution is to solder a piece of nickel silver or, better still, an old relay contact to the two points which contact the battery. Corrosion will still occur on the supporting springs but since it no longer affects the contact resistance it is relatively unimportant. The corrosion is mainly surface corrosion and has little effect on the strength of the springs.

Alkaline cells use a central powdered-zinc negative electrode with an alkaline electrolyte (potassium hydroxide) surrounded by a manganese-dioxide positive electrode. They are a considerable improvement on the basic zinc carbon cell in almost every respect. Unfortunately they are also more expensive. Some of this increase in cost can be offset against the improved performance when delivering high currents for long uninterrupted periods, but the extensively advertised 6:1 increase in life needs to be examined with some care. This degree of improvement only occurs when compared with the basic zinc carbon cell operated continuously on a high current load i.e. a thoroughly unsuitable load. If the comparison is with a zinc chloride/HP cell or if the zinc carbon cell is allowed rest periods on no-load the difference is considerably smaller. On moderate loads the improvement is about 3:1 and on low-current loads it is less than 2:1. However on extremely low-current loads the advantage returns because of the better shelf-life of the alkaline system.

Alkaline cells also suffer loss of capacity if discharged at high rates although not as severely as zinc carbon and HP cells. Table 7.2 lists the nominal capacities of the common cell types. However, these figures are the capacity which is delivered at low-discharge rates — typically 100 hours to full discharge. At discharge times of a few hours, typical of many small motor applications, the useful capacity will be less than half this value.

When choosing battery size it makes good sense to err on the large side. The larger sizes cost less per mAh and, because the discharge times are then longer, deliver a much larger fraction of their nominal capacity. A D size cell replacing an AA size will deliver well over four times the useful operating time on a typical small motor load.

7.2.2 Mercury cells
Mercury cells are mainly encountered as small button cells in deaf aids. They are easily produced in small sizes and deliver a little more capacity per unit volume than alkaline cells. They are characterised by a low (1.35V) and extremely constant cell voltage with a flat discharge characteristic which only falls to 1.1V at the end of life. Larger sizes are used in military and some professional application which take advantage of their wide operating temperature range and stable discharge characteristics but they are too expensive for general-purpose use.

7.2.3 Silver-oxide cells
Silver-oxide cells are again produced as

Table 7.2 Alkaline cells

Size	Voltage	Diameter mm	Length mm	Capacity mAH
N	1.5	12	29	800
AAA	1.5	10.5	44.5	1,100
AA	1.5	14.6	50	2,700
C	1.5	26	50	7,700
D	1.5	34.2	61.2	18,000
PP3	9	16 × 25	48	550
MN21	12	10	28	33

small button cells and dominate the watch battery market. They have about the same energy density as a mercury cell but with the key advantage that they have a flat discharge characteristic with the output voltage remaining above 1.4V for most of the useful life. This voltage is high enough to allow the electronics to operate from a single cell.

7.2.4 Lithium cells
Lithium cells are a more recent development and are characterised by a very high initial voltage (3V to 3.5V depending on type) together with exceptionally long shelf-life. For the same cell size the ampére/hour capacity is similar to silver oxide but since the output voltage is about doubled the watt/hour capacity is proportionately higher. For specialised professional use there is a wide range of lithium cell types based on several chemical systems. However the readily available commercial types are limited to a few cylindrical cell types for primarily photographic applications and a range of button cells for calculator and computer memory backup purposes.

The photographic types use a lithium/manganese dioxide system and are optimised for short bursts of high current drain to drive camera motors and electronic flash systems. The button cell types are mainly intended for low-current drain applications with battery life measured in months or years.

When lithium cells are used in conjunction with other power sources, extreme care must be taken to prevent reverse current flowing into the cell. Any current trying to charge up the cell will cause release of gas which is unable to escape and, over a period of time, pressure will build up and cause the cell to burst. The problem is acute in computer memory backup applications where the reverse current hazard may be present for months or years. The reverse current isolation provided by ordinary silicon signal diodes is not good enough and selected low-leakage types are necessary.

7.3 Rechargeable cells

7.3.1 General
Rechargeable cells are described in terms of their ampére-hour capacity and this is usually abbreviated to the single letter "C". For example a cell discharged at C/10 rate would be fully discharged in 10 hours, one discharged at 5C rate would be discharged in about $\frac{1}{5}$ hours = 12 minutes.

The word "about" is used in the high rate discharge example because the useful ampére-hour capacity of a cell reduces at high rates. Cells designed for high peak currents may deliver ten or eleven minutes at 5C rate. Other cell types may only deliver output for a minute or so and some may be unable to deliver the current corresponding to 5C rate even when directly short circuited.

The same nomenclature is used for specifying the recharging current e.g. a 5 Ah cell charged at C/10 rate would charge at a rate of 5/10 A = 500mA. In an ideal world this would fully recharge a discharged cell in 10 hours. However, the efficiency of the recharging process is rarely higher than 70% so that about fourteen hours would be necessary.

7.3.2 Recharged primary cells
Most primary cells can be recharged to some extent but the results are seldom worth the effort. Only a fraction of the original capacity can be restored and there is a danger that the cells may leak or burst. Most manufacturers state either that their primary cells are not rechargeable or expressly forbid it. If you should

be tempted to experiment along these lines be very sure that a leaking or bursting battery cannot harm you or anyone else.

There are a few primary cell battery chargers on the market and while these can undoubtedly significantly extend the useful life of suitable primary batteries they are not miracle 'new batteries for old' machines. They can restore a partially discharged battery to near its fully charged state several times but the self-discharge rate of the recharged battery is much higher than the original. To be really effective less than one-third of the battery capacity should be used before recharging and the recovered charge of the revived battery used within a day or so. Less of the battery capacity is recovered on each cycle and a fully discharged or an old partially discharged battery will usually not respond at all. Alkaline batteries respond somewhat better than Leclanche and HP types.

7.3.3 NiCad cells

Sealed nickel cadmium rechargeable cells (NiCads for short) are probably the most generally useful rechargeable cell. They can be charged and discharged at very high rates and are very tolerant to neglect and occasional misuse. They are available in two main types — a range of cylindrical cells suitable for general purpose and high discharge rate use, and a range of mass plate button cells for low-current applications and where long-term charge retention is important.

Data on commonly available types is given in Tables 7.3 and 7.4. The capacities listed are approximate as they vary slightly between manufacturers and between particular cell types. However, they are a good general guide — in the case of the very popular C size cell the professional types range between 1.4Ah and 2.2Ah but are mostly 1.8Ah or 2.0Ah. Domestic types sold in supermarkets and similar outlets are a cheaper version and are almost universally 1.2Ah. Note the enormous difference in capacity between professional and domestic D size cells. The domestic D size is simply a sub-C size cell mounted inside the larger D size case. They are easily distinguished because nickel and cadmium are heavy metals and the domestic D cell is less than half the weight of the professional version.

Note that although many of the NiCad cell sizes are identical to alkaline primary

Table 7.3 Cylindrical NiCad cells

Size	Diameter mm	Length mm	Capacity mAh	16 hour charge rate mA
AAA	10.5	44.5	220	22
AA	14.6	50	500	50
sub-C	22.7	42.1	1,200	120
C (professional)	26	50	1,800	180
C (domestic)	26	50	1,200	120
D (professional)	34.2	61.2	4,000	400
D (domestic)	34.2	61.2	1,200	120
PP3	16 × 25	480	120	12

Table 7.4 Button NiCad cells

Size	Diameter mm	Length mm	Capacity mAh	16 hour charge rate mA
*	11.7	5.55	30	3
*	15.8	6.25	75	8
*	22.6	4.8	120	12
*	25.5	6.0	190	19
*	25.1	9	280	28
*	35	10	600	60

*There are no standard designations for button cells but these are commonly encountered sizes.

cell types, NiCads store less than one-quarter of the energy available from the same size alkaline primary cell. This is the penalty for rechargeability.

On the plus side, NiCads are very much better at high discharge rates. One to two amps is about the maximum useful peak discharge rate for a C size alkaline, but the same size NiCad will cheerfully deliver fifty amps for a short and glorious discharge cycle!

The cylindrical cells use positive and negative electrodes in the form of long thin strips which are interleaved with a thin porous plastic separator. This is then rolled up into a sort of 'Swiss roll' and sealed inside a nickel-plated steel outer case. Enough electrolyte is added to saturate the separator. There is no free liquid electrolyte so the cell is not position-sensitive and can operate any side up. This form of construction packs a large electrode area into a small space. The cell design has an exceptionally low internal resistance — a few thousandths of an ohm so that it is capable of delivering very large currents.

The mass plate button cells achieve the necessary large surface area needed by using electrodes in the form of porous sintered plates. Both positive and negative electrodes are porous discs of sintered metal powder which provide such a large internal surface area that only two or three discs are required to make up a cell. This construction has a lower self-discharge rate than cylindrical cells and stores about the same energy per unit volume. However, the internal resistance is considerably higher and these cells are not suitable for high discharge rates.

For routine low-rate recharging, chargers are designed to deliver a fixed current that recharges a cell in about sixteen hours. This charging rate is the C/10 current i.e. the current that would fully discharge the cell in ten hours. Low-rate charging is not very efficient. About one-third of the charge energy is converted into heat so that at least sixteen hours at C/10 is needed for a full recharge. The great advantage of low-rate recharging is that the cell can withstand prolonged overcharging for days or even weeks at this rate.

Prolonged overcharging does have a small effect on cell life and C/10 should not be used for float charge operation i.e. for standby operation where it is permanently connected to a charging source to ensure that it always remains fully charged. Float charge rate should be limited to about C/100 for button cells or C/50 for cylindrical cells. Typical performance for a button cell operated under these conditions is a life of about three years. With the exception of a few specialised types, cylindrical cells are not really suitable for float charge operation because they suffer from 'memory effect' (see later) and their higher self-discharge rate necessitates a larger float charge current which has an adverse effect on life.

The sub-C size cell is popular in high discharge rate and fast charge applications. Typical uses are portable power tools, model electric racing cars, camcorders and similar devices. Premium versions of these cells are available in matched sets which are capable of delivering most of their rated capacity at 5C rate i.e. fully discharged in 10 minutes!

Cell life in normal charge/discharge operation depends on charge and discharge conditions. The most unfavourable cycle is deep discharge followed by prolonged overcharge and cell life may be no more than a few hundred cycles. Partial discharge, followed by no more than the recommended recharge time, gives the best life and several thousand cycles may be possible. Unfortunately, if the cell is always treated gently, it becomes 'lazy' after many cycles (memory effect) and exhibits a temporary loss of capacity. The cell is not usually harmed by this and full capacity can be restored by subjecting it to several deep discharge, full recharge cycles. Memory effect is mainly encountered in cylindrical cells and the temporary loss of capacity is often quite small (less than 20% with some types). Unless it is essential that the cells be kept in maximum capacity condition it may be better to accept some loss of capacity rather than reduce cell life by frequent deep discharge cycles.

Sintered plate button cells are designed to be especially suitable for shallow discharge battery backup applications and do not suffer from significant memory effect.

As long as the correct charging conditions are observed the cylindrical cell types can be recharged more rapidly, but the recharging conditions must be more closely controlled. Table 7.5 shows charge time restrictions for charging rates of up to C. For charge rates of up to C/5,

Table 7.5 NiCad charge rates

Charge time hours	Charge rate	Remarks
16	C/10	Normal low rate charge May continue for several days
6	C/5	Quick charge Limit charge duration to less than 7 hours
1	C/1.25	Fast charge Batteries must be discharged to 1V per cell before start of charge Charge must be terminated at 1 hour with cell 80% charged. Top up at C/10 or lower rate

provided the cell is at least partially discharged and the charge is terminated promptly, it is not essential to start with a fully discharged cell. Rates higher than C/5 must start with fully discharged cells. For the very highest charge rates only partial recharge is advisable.

Cells subjected to very rapid discharge can get quite hot. They must be allowed to cool down before any form of rapid charging is attempted.

The above comments apply to simple chargers using a fixed current for a known time. More sophisticated chargers can recharge cells in only an hour or so even if the previous state of charge is unknown. These measure the voltage and current characteristics and/or the cell temperature during charge and modify the charge rate accordingly.

Button cells should **NOT** be rapid charged. This is partly because of their higher internal resistance but also because this type of cell is not fitted with a pressure relief valve. The first indication of excessive charge rate is the cell starting to expand into a near spherical shape and, in extreme cases, bursting!

As single cells, NiCads can be stored in any state of charge. They are not harmed by being completely discharged and are normally shipped in this condition. However more care is needed with a stack of cells forming a complete battery.

In a battery of cells, one cell may have a slightly lower capacity and, on load, will completely discharge before the remainder. The remaining good cells will continue to supply current to the load. This current will flow through the completely discharged cell and try to charge it up with reversed polarity (see Fig. 7.1). This has two adverse effects. First, the reverse charge delays the effective start of the next charge cycle for that cell so that even less capacity is available for

the next discharge cycle. Secondly, during reverse charge, gas may be generated in the cell at excessive rate and vented through the safety valve located under the bump that forms the positive contact on the cell. Any gas that is vented in this way cannot be recombined to form the water that the cell needs and so results in a small permanent loss of capacity.

There are several ways to avoid this problem. The simplest method is to never discharge a battery below an average of 1V per cell. With small numbers of reasonably well-matched cells this average voltage is high enough to ensure that no cell becomes reverse-biased. Once cell voltage has reached 1V it is falling rapidly and continuing to a lower voltage delivers little extra capacity. If a battery is inadvertently discharged beyond this point no great harm is done, but follow it by a prolonged low-rate recharge to ensure that the weakest cell receives a full recharge. An alternative solution is to

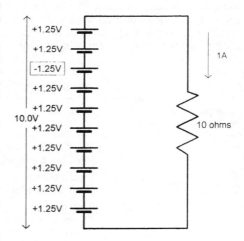

Fig. 7.1 *Weak cell polarity reversal during deep discharge*

connect a silicon diode in parallel with each cell (see Fig. 7.2). In normal operation this diode is reverse-biased by the cell voltage and does not conduct. If the cell completely discharges and the current supplied by the remaining cells tries to force a reverse voltage across it, the diode conducts and prevents the reverse voltage across the cell rising beyond about 0.8V. This is a good method for load currents of up to a few amps but beyond this the diodes become a significant size and cost item.

For the extremely high rates of discharge typical of model car racing neither method is very suitable. The batteries are being used at the limit of their power capacity and normal on-load cell voltages will often be below 1V. Diode protection is still possible but needs expensive diodes. The normal solution is to buy the batteries as a computer-matched set of cells or as a factory-matched assembly.

A lot depends on the number of cells in the battery. A cut-off point corresponding to an average of 1V/cell is pretty safe up to about 8 cells in series. Reverse polarity is still possible on a badly mismatched cell but is unlikely to be bad enough to lead to permanent damage provided that the battery (particularly including the weak cell) is *fully* recharged after each deep discharge. Failure to ensure this will result in deeper reverse charge on the weak cell on each cycle and early failure.

At 10 cells in series, as shown in Figure 7.1, it becomes easier for mismatched cells to seriously reverse drive the weak cell (third from the top) and, unless well-matched cells are used, it is advisable to limit the reverse voltage by diode protection as shown in Figure 7.2. At 20 cells in series (the normal number for military 24V to 28V supplies) the problem can be acute and it is common practice to include diode protection, even on initially well-matched cellpacks. Diode protection does not entirely eliminate reversed polarity but limits the maximum reverse voltage to a safe level.

If the very highest charge capacity is needed the battery should be discharged to less than 1V per cell and then rapid charged followed by a lower rate top-up charge. A car headlamp bulb is a convenient discharge load but it is very important that the discharge process is not taken far enough to reverse charge the weakest cell. A much better and safer way of rapid discharging a NiCad battery is to connect a low-value discharge resistor across *each* cell. With this method the cells can be discharged as rapidly, and completely, as desired and there is no possibility that any cell can become reverse-biased.

Typical charge and discharge curves are shown in Figures 7.3 and 7.4. On charge, the voltage across an initially

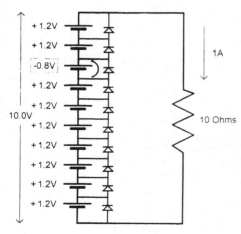

Fig. 7.2 *Diode protection against polarity reversal*

discharged cell rises rapidly to a bit less than 1.4V and then more slowly to just over 1.5V. The cell voltage is strongly dependent on temperature as can be seen from the three curves in Figure 7.2. When charged at room temperature, the cell is nearly fully charged at about eleven hours and most of the charge power now appears as heat instead of chemical change. The increase in cell temperature reduces the cell voltage so that further charging results in a small *drop* in cell voltage. The size and shape of this voltage peak is very dependent on charge rate, temperature and cell design and, at low charging rates may be completely absent. At high charge rates the effect is more pronounced and some of the more expensive chargers use the voltage drop after the peak to switch from a high-rate charge to a lower top-up rate.

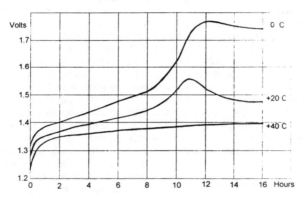

Fig. 7.3 *Charge characteristics*

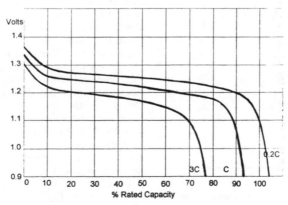

Fig. 7.4 *Discharge characteristics*

When charge current is removed the open circuit voltage declines within minutes to about 1.4V and down to 1.35V in a few hours. There is not a direct relation between open circuit voltage and state of charge and little charge is lost in this initial drop. The self-discharge rate of cylindrical NiCads varies between samples and is quite high in the first few days. It is usually safe to rely on about half capacity remaining after one month and perhaps one-tenth after three months. Good new cells can do considerably better than this. Figure 7.4 shows typical discharge characteristics for a discharge rates of C/5 (this is the normal discharge rate for rated capacity), and also for 1C and 3C.

At C/5 rates the cell voltage drops quickly to 1.25V and drops slowly to 1.2V when about 10% of the charge remains. It then drops fairly rapidly to the end point of its rated capacity at 1.1V. Beyond this point the voltage falls steeply and there is little point in trying to continue below 1.0V per cell.

At very high discharge rates (more than 1C) the plateau is less marked and the output voltage and also the effective capacity is reduced by the internal resistance of the cell.

Pressure contacts are not satisfactory at very high discharge rates and it is necessary to solder wires directly to the cells. This is **NOT** recommended by the manufacturers and will invalidate any warranty. However, there may be no alternative and if carried out carefully is usually successful. The trick is to thoroughly clean the surface and use a very hot soldering iron so that the whole operation is complete in a few seconds before the main body of the cell has a chance to heat up.

The ideal iron for the job is a 40W temperature controlled iron set to its maximum temperature. This is an expensive luxury so the alternative is an ordinary 40W iron or an old-fashioned copper iron heated by a blowlamp. The small 15W irons used for electronic assembly are not man enough for the job.

Work with the surface uppermost and scrape the joint area bright shining clean. Wet the area with a single drop of Bakers Soldering Fluid (this is a powerful liquid acidic flux — most paste- and resin-based fluxes are not sufficiently aggressive for this job). Melt a small blob of solder on the end of the iron and bring it into firm contact with the joint area. Within a second or two at most the solder should flow and tin the cleaned area. Allow to cool and wash off the flux residue with water/detergent.

If the solder fails to fully wet the surface at first attempt, *do not* scrub away with the soldering iron but immediately remove the soldering iron and allow the cell to *completely* cool down to room temperature. Repeat the attempt with a fresh and more thoroughly cleaned surface.

Now using ordinary resin-cored solder, tin the end of the connecting wire. Make the joint by pressing the tinned wire into the tinned area with the soldering iron and, at the same time, feeding in a little more resin cored solder to complete the thermal contact. As soon as the whole solder blob melts remove the iron and you should be left with a near-perfect joint. Artists with a soldering iron can complete this process in one step using only resin cored solder, but the two-step process using Bakers fluid is kinder to the cell and gives consistently good results. If you are newcomer to this sort of soldering, first practise on a piece of nickel plated or bare mild steel.

NiCads have two main failure modes. The first and most common is progressive loss of capacity. This is the normal wearout mechanism of the cell and is accelerated by excessive overcharging or frequent deep discharge/recharge cycles. Although the cell is fairly well sealed it is not a hermetic seal and there is a very slow loss of electrolyte mainly in the form of water vapour. A very small amount of the potassium hydroxide (caustic potash) active constituent may also migrate and this is often seen as a white powdery deposit around the positive contact. Fortunately, caustic potash rapidly combines with carbon dioxide from the atmosphere and the white deposit is not the caustic electrolyte but the comparatively inert potassium carbonate.

Much of the capacity of a cell that has been lost by drying out can be recovered by drilling a small hole in the case and injecting a measured amount of water. This is a desperate move and not recommended (see 'Nickel-cadmium cells' by K. C. Johnson in *Wireless World*, February 1977, pp. 47—8). If you should be tempted to experiment don't forget that there may be a high pressure inside the cell (the pressure release valve hidden under the positive contact doesn't open until the internal pressure reaches about 200 psi!) and some of the caustic electrolyte could be released.

As mentioned earlier, cells can suffer a temporary loss of capacity if subjected to a long series of shallow discharge/ slow recharge cycles. This is caused by the slow formation of relatively large crystals in the active area instead of the normal fine structure and this results in a loss of active surface area. Occasionally these crystals can take the form of needle-like structures which penetrate the inter-electrode spacer and short circuit the cell. A cell with this fault will register zero volts no matter how long it is charged at any normal charging rate.

This type of failure is common if the cell is frequently low-rate charged for very long periods with little or no inter-mediate discharge operation. Lead—acid cells thrive on this sort of treatment but NiCads prefer to be at least partially discharged before being subjected to a full low-rate recharge cycle.

Cells that have failed in this way can often be recovered by subjecting them to a short high-current pulse to burn out the offending needles. The current must be high enough to burn out the needles but once the short circuit has been cleared the voltage must not rise high enough to damage the inter-electrode spacer. A convenient source is two series-connected fully charged NiCads of similar or larger capacity.

Put the faulty cell on charge in the normal way with a meter across it to monitor cell voltage. Apply the high-current source (+ve to +ve and −ve to −ve) for about one second − it should raise the cell voltage to about 2V. The cell voltage will drop back when the high current source is removed but the normal charge current should bring it up to the normal voltage range within the next minute or two. If the cell voltage remains near zero repeat the treatment. Two or three pulses are enough to clear most cells but a few may remain obstinately stuck at zero. Increasing to three or four series-connected cells may do the trick but the chances are not good. Don't prolong the current pulse − if the needle is going to burn out it will do so in the first second or so. Longer applications will cause excessive general heating in both the faulty cell and the current source cells.

Once the short circuit is cleared follow by several full charge/deep discharge cycles to restore the normal crystal structure.

7.3.4 Nickel-hydride (NiMH) cells

These are a more recent development and are a popular replacement for NiCads in premium applications such as portable computers and some classes of military equipment.

They are manufactured in similar sizes and shapes to the major NiCad sizes. The cell voltage characteristic over the discharge cycle is almost identical to NiCad cells but, for a given cell size, the nickel-hydride cell under optimum conditions delivers 50% to 100% additional capacity.

However, this performance can only be achieved at fairly light loads — C/3 rate or less i.e. time for complete discharge 3 hours or longer. If discharged at higher rates the internal losses are excessive which leads to self heating and reduced output voltage and it is no longer better than a standard NiCad.

The self-discharge rate is about three times worse than NiCads so that the main field of application for these cells is for use on moderate loads without long intervals between recharging and subsequent use.

The cells can be constant current low rate (15 hours) recharged on standard NiCad chargers. They can also be fast charged (1 hour) but special chargers are needed because the end of charge drop in cell voltage which is normally used to terminate the fast charge is smaller than the NiCad end of charge voltage drop

At present these are considerably more expensive than NiCads although this is reducing as manufacturing quantities increase.

7.3.5 Lithium-ion cells

This is a recent entry into the rechargeable field and has an even higher energy density than NiMH cells. They cannot be used as direct replacements for NiCad/ NiMH systems because of the very different discharge characteristics. The initial voltage at full charge is about 4V and this falls steadily as it discharges to the discharge cut-off point of a little before 2.5V.

The high-rate discharge performance is rather better than NiMH cells and they can be rapid recharged from a constant voltage source in one to two hours. However, both charge and discharge conditions must be extremely tightly controlled. The cells are damaged if overcharge allows the cell voltage to rise beyond 4.2V and may fail completely if discharged to below 2.5V.

7.3.6 Lead—acid rechargeable cells

Perhaps the most familiar lead—acid batteries are the 12V starter batteries fitted to almost every car. At present most of these are unsealed with the electrolyte accessible for topping up, but fully sealed zero maintenance types are now fitted in many cars. For most small workshop projects the smaller fully sealed batteries are the most useful — a range of these is shown in Figure 7.5.

The rectangular block type is most commonly encountered and is manufactured in sizes ranging from less than 1Ah to more than 50Ah. The electrodes consist of lead cast into rectangular plates which carry the active material. Metallic lead is the main current carrying conductor but the active material takes the form of a paste pressed into the lead matrix which covers the working area of the positive and negative plates. Typically three to eleven alternate positive and negative plates are interleaved and separated by porous plastic spacers to form a cell. The porous spacers are saturated with electrolyte which is dilute sulphuric acid. There is no free liquid acid and the electrolyte is always retained within the active volume. This makes it possible to

Fig. 7.5 *Sealed lead–acid batteries*

operate the cell any way up.

A cylindrical form of construction is favoured by Cyclon. This uses the 'Swiss roll' electrode configuration mechanically similar to cylindrical NiCads. This results in extremely low internal resistance – the 5Ah cell is rated to deliver 200A peak current!

Although often used in similar applications, the characteristics of sealed lead–acid cells are very different from their NiCad counterparts and this must be taken into account in the management of their charge/discharge cycles.

The useful life of lead–acid batteries is degraded by frequent deep discharges. Lead–acid batteries should never normally be discharged to below 1.8V per cell – below this voltage part of the active material in the plates start to convert to a form of lead sulphate which reduces the ability of the cell to accept charge with consequent loss of capacity i.e. the cell becomes 'sulphated'. (*Note* 1.8V is for a cell that is being discharged at normal or light loads that could be supplied for 10 or more hours. For very heavy loads, such as a traction motor load that would discharge the cell in one hour, the internal resistance of the cell may drop

the apparent terminal voltage by an additional 0.1V or 0.2V. In this case an endpoint of 1.7V or 1.6V – 10.2V or 9.6V for a 12V battery – is tolerable although not recommended. The terminal voltage will, of course, rise to nearly 2V as soon as the motor load is removed.) In fact there is little energy left below 2V per cell and if the discharge is terminated at this level and the battery recharged without undue delay, battery lives of well over a 1000 charge/discharge cycles are possible. With frequent discharges to 1.8V per cell and lower, the plate structure starts to disintegrate and the life will be reduced to a few hundred charge/discharge cycles.

Because of this, unlike NiCads, lead–acid batteries should *not* be completely discharged. They do not suffer from the NiCad memory effect and should be recharged as soon as possible after use and, in any event, before the voltage falls to 1.8V per cell. The self-discharge rate of the sealed batteries is much better than that of the open car-type batteries but it is still advisable to give a top-up charge at least every six months to batteries stored for long periods.

If you are buying lead–acid batteries

on the surplus market, take a voltmeter with you to measure the open circuit voltage. On no-load, a fully discharged battery will read just under 2V per cell. A battery that has dropped to 1.7V per cell has started to sulphate up (see later), but is probably recoverable. Below 1.7V per cell there is likely to be permanent loss of capacity. This is quite different to the behaviour of NiCads which can be happily stored, discharged to zero volts!

Correct charging is very important. Ideally lead−acid batteries should be charged at a fixed voltage of 2.4V per cell (14.4V for a 12V battery). The current at the start of charge should not exceed Ah/4 (i.e. 2½ amps for a 10Ah battery) but this is not a problem with most small chargers. If the current is excessive, then a short length of electric fire element or florists' iron wire in the charger output will bring the initial current down to safe limits.

A more serious problem is the charging voltage at the *end* of charge. Purpose-built commercial chargers have a built-in voltage regulator which keeps the output voltage constant throughout the charging period. At 2.4V per cell a battery will be about fully charged in about 4 to 6 hours, and the residual charging current will have dropped below Ah/20 (½ amp for a 10Ah cell). This adds little additional charge because most of this current is not true charging current but is starting to break down the water in the electrolyte into oxygen and hydrogen (i.e. gassing). At this current level the chemical system in the cell can recombine the gases into their original water form. The composition of the electrolyte does not change and there is little increase in the internal pressure in the cell.

However if a typical DIY car battery charger or a NiCad charger is used the voltage will not be closely regulated and,

as the current drawn by the battery drops towards the end of charge, the charging voltage rises well above the 2.4V per cell level. The gassing rate increases rapidly as the cell voltage increases and, at somewhere between 2.5V and 2.8V per cell (15.0V to 16.8V for a 12V battery) exceeds the rate at which the cell chemistry can recombine it to water. The gas pressure in the cell rises and the excess gas is vented through the safety vent. There is very little excess water in the electrolyte and this lost water represents a permanent loss of capacity. In extreme cases prolonged over-voltage charging can dry out a cell almost completely.

Apart from buying an expensive voltage-regulated charger it is not too easy to completely avoid this problem. Fortunately it results in steady degradation rather than catastrophic failure, so exceeding the voltage or current limits simply results in reduced capacity and shorter life rather than complete failure. The damage is reduced if you use a small car battery-type charger and do not leave the battery on charge for unnecessarily long periods.

The readily available NiCad chargers are the worst sort as these are designed to deliver an approximately constant current — the exact opposite of the lead−acid constant voltage requirement (a constant current charger delivers the same current all the time and the current does not drop much as the battery voltage increases). The only way to safely use these is to choose a current setting, or to add additional resistance, to ensure that the final current is less than Ah/20. The charging current efficiency at these low charging rates (Ah/20 is not much more than a trickle-charge maintenance rate) is very poor and more than 30 hours will be needed for full recharge. However, it has the advantage that the charging time is not critical and even 60 hours charging is unlikely to do much harm.

If in doubt check the final voltage with a good meter — use a digital voltmeter if possible because few analog voltmeters can be relied upon to much better than 5% and this is not good enough. Always check the charger output voltage at or near the end of charge **WITH A BATTERY CONNECTED**. The output of the charger is not smooth DC but has a large ripple component and this introduces large errors unless a battery is connected.

DIY car battery chargers are a better bet but, unless you are using a small charger to charge a rather large battery, the output voltage at the end of charge will be well over the 15V permissible for a sealed 12V battery. The simplest way of dealing with this is to connect a small voltage regulator between the charger and the battery as shown in Figure 7.6.

The LM317T is good for charging currents of up 1½A and has built-in thermal overload protection. It will get quite hot if supplying anything like its full current rating so the mounting tab should be securely bolted to a suitable heatsink or, via an insulating mounting kit, to the main metalwork of the charger. Unlike the popular 78xx and 79xx series regulators, the LM317T tab is connected to the output terminal of the regulator.

At the end of charge the battery should be disconnected from the charging regulator to prevent it slowly discharging through R1 and R2.

The above comments on charging difficulties apply only to completely sealed lead—acid batteries. There is no problem with unsealed car batteries because any water lost by gassing is easily replaced by topping up with distilled water.

The appropriate charge voltage is 2.4V per cell for cells in cyclic use i.e. cells

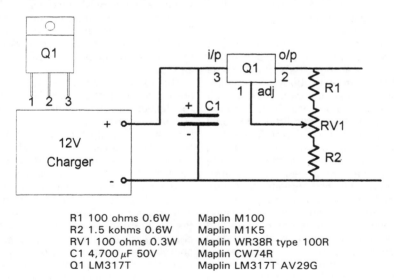

R1 100 ohms 0.6W   Maplin M100
R2 1.5 kohms 0.6W   Maplin M1K5
RV1 100 ohms 0.3W   Maplin WR38R type 100R
C1 4,700 μF 50V   Maplin CW74R
Q1 LM317T   Maplin LM317T AV29G

Fig. 7.6 *Charging regulator*

that are frequently first charged and then fully or partly discharged. It should be reduced to 2.3V per cell for batteries which are kept on continuous float charge for standby applications.

Another problem with both sealed and unsealed batteries is sulphating of the plates. This occurs if the battery is allowed to stand for long periods in a low state of charge. A white film of lead sulphate forms on the surface of the plates and since this is a semi insulator the internal resistance of the cell eventually rises to the point where it can no longer deliver or accept charge. This should not occur if you recharge soon after use and give a top-up charge at regular intervals to batteries that are stored for long periods.

However, it is easy to forget top-up charges and you may be faced with a sulphated battery. If it is only slightly sulphated it will still accept charge at a reduced rate from a normal charger and deliver half or more of its rated capacity. Two or three charge and discharge cycles should clear this and restore most of the rated capacity. Because of the reduced charge current, the charge time should be longer than normal. Discharge should be into a lamp load that would discharge a reasonably good battery in a few hours — perhaps three or four. For the first one or two discharges you won't know how long the charge will last and you don't want to further degrade the battery by over-discharging. This is where the lamp comes in — put it where you can see it and end the discharge as soon as it starts to dim.

If the battery is badly sulphated, the output voltage of a standard charger may be not be enough to break through the sulphate film and the battery will not accept any charge at all. Provided you have the right sort of power supply (up to 30V for a 12V battery) there is still a fair

chance of recovery although it is a long process. The following section is based on the method recommended by Yuasa.

Sulphation can be broken down by constant current charging for up to twelve hours with the current limited to not more than C/10. With a 12V sulphated battery as much as 30V may be needed to reach C/10. As the sulphation is broken down the battery voltage will start to drop and, as soon as it drops to 15V, constant current charging should be discontinued and charging continued for 5 hours on a normal constant voltage charger set to 15V output.

It's important that the current is limited and the battery is not allowed to get warm during the first high voltage phase. This is because the sulphate film does not break down uniformly — at first, most of the current that flows is concentrated into a few isolated spots where the sulphate film has been penetrated. Although the total power into the battery is small it is all concentrated into these small areas and, if the current is too high, these can easily get hot enough to permanently damage the plastic separating membrane — this can occur while the battery as a whole is barely warm.

If the battery was badly sulphated this may only result in partial recovery. However, two or three full charge/full discharge cycles should recover most of the original capacity.

It is difficult to obtain 30V constant current chargers, but a good enough approximation is to connect a fully charged 12V car battery and a resistor in series with an ordinary car battery charger as shown in Figure 7.7. The addition of the capacitor plus the car battery brings the total output up to about 30V. The resistor value is not critical but should be chosen to limit the current to between C/20 and C/10 when the sulphated battery has

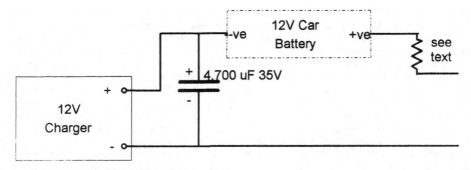

Fig. 7.7 *High-voltage charging rig*

started to recover and the voltage across it has fallen to 15V.

With most car batteries there is the additional freedom that it is possible to change the electrolyte. The Chloride Electrical Storage Co. recommend that the acid electrolyte be replaced by distilled water and the battery given a prolonged charge until the specific gravity rises to 1.100 – 1.150. The battery is then refilled with fresh distilled water and charging continued until the SG ceases to rise and remains steady. The battery is then refilled with fresh acid SG 1.250 and given a normal charge.

Both of these processes are for recovery of sulphated batteries that are otherwise in reasonably good condition. There is no point in trying them on cells whose plates are badly buckled or starting to disintegrate.

CHAPTER 8

Associated Components

8.1 General

This chapter is limited to just a few essential components and test equipment used in conjunction with motors and their control systems (a wider range is covered in the earlier book *Electric Motors* published by Nexus Special Interests).

8.2 Capacitors

8.2.1 General

Capacitors consist of two metal electrodes separated by a thin layer of insulating material. They are a bit like rechargeable batteries that can be charged and discharged with extraordinary rapidity — fractions of a millionth of a second. By rechargeable battery standards their energy storage capacity is rather small. Even small rechargeables can store watt/hours of energy. The start and run capacitors used with motors store only watt/seconds of energy but since this charge is circulating in and out of the capacitor one hundred times a second (i.e. once each negative and once each positive half cycle of the 50Hz mains) the circulating power is hundreds of watts. This is usually called volt amps to differentiate it from watts because very little of this power is dissipated within

the capacitor — usually less than 1%.

The fundamental unit of capacitance is the farad but this is too large for everyday use so capacitors are mostly marked in μF which are microfarads i.e. millionths of a farad, or in pF which are picofarads i.e. millionths of a μF. Many different types are used in electronic applications and in an enormous range of values. A computer may use capacitor values ranging from a few pF in the high-speed logic to several hundred thousand μF for memory backup.

Capacitors used in connection with motors cover a much smaller range — about 1 μF to 500 μF for start and run capacitors for induction motors and 0.001 μF (= 1000 pF) to 0.1 μf for interference suppression on commutator motors.

Quite large currents flow through capacitors in motor circuits but this is because the alternating current supply is causing charge to flow in and out of the capacitor at supply frequency. In any DC circuit the capacitor charges up almost instantly to the supply voltage and that ends the current flow. If a large capacitor is checked with an ohmmeter there is an initial kick as the capacitor charges up to the voltage of the battery in the ohmmeter, but after that, no further current can flow so, the meter settles down to read open circuit.

Capacitors can be connected in parallel or in series to make up non-standard values (see Fig. 8.1).

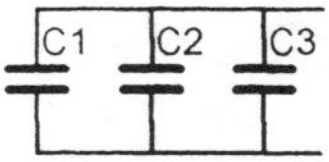

Parallel
$$C = C1 + C2 + C3$$

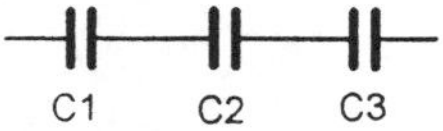

Series
$$C = \frac{1}{1/C1 + 1/C2 + 1/C3}$$

Two equal capacitors in series - capacity = C/2
Three equal capacitors in series - capacity = C/3 etc.

Fig. 8.1 *Series and parallel connection*

If a capacitor is disconnected from a circuit it will remain charged up to the voltage that existed on it at the moment of disconnection. This charge may take hours or days to leak away and if it is a large capacitor charged to a high voltage this can give an unpleasant shock to unwary fingers. Run and start capacitors operating in mains voltage circuits come into this category and should be treated with respect. It is good practice to permanently connect a slow discharge resistor across the terminations of this type of capacitor (100K ohms 1W is a good value) and this will reduce the discharge time to a few seconds or tens of seconds. If in doubt, discharge the capacitor through a low value resistor − about 100 ohms. Don't use a screwdriver to short circuit large high voltage capacitors − the resultant explosive discharge is not good for the capacitor or the screwdriver!

8.2.2 'Run' capacitors
These are capacitors that are in circuit the whole of the time that the motor is running. They must be rated for continuous operation at the full AC mains voltage. Types designed for AC working will include the letters AC as part of the marking giving capacity and working voltage e.g. 7.5 μF 250V AC.

Originally this type of capacitor used oil-impregnated paper as the insulating layer but modern capacitors use a thin plastic film usually Mylar, polycarbonate or polypropylene. The electrodes take the form of long strips of aluminium foil or, in some cases, an extremely thin coating of zinc or aluminium deposited directly onto the insulating film. Capacitors of the size needed for motor capacitors need many square feet of electrode area. The very thin sandwich formed by the electrodes and insulating layers is rolled up on itself, either as a straightforward 'Swiss roll' for cylindrical capacitors or on a thin flat former to produce an approximately oval section suitable for stacking in rectangular housings.

This sort of capacitor is not too common in electronic equipment. They are most often encountered as power factor correction capacitors in industrial fluorescent lamp fittings and they are, of course, associated with some types of induction motor. The types used in fluorescent fitting are rarely larger than 8 μF so two or more are necessary for most motor applications. A selection of capacitors used with fluorescent lamp fittings is shown in Figure 8.2

They are also stocked as spares for the earlier types of automatic washing machines.

Paper and film capacitors with very similar internal construction to the AC-rated types are used in military and industrial electronic equipment. These

Fig. 8.2 *Fluorescent lamp capacitors*

are rated for DC working but with sufficient derating are fine for use on AC. For 240V AC working the DC rating must be at least 350V and preferably more. Capacitors with values large enough for motor work are usually housed in rectangular steel cans. Examples are shown in Figure 8.3.

If you use any capacitor which is not marked with an AC rating, check that the terminations are not polarity marked in any way. Red and black or plus and minus coding are sure indicators that it is a electrolytic capacitor for use on DC

Fig. 8.3 *Steel can paper and film capacitors*

supplies (see below) and *not* suitable for use as run capacitors.

8.2.3 'Start' capacitors

Typically two to five times more capacitance is needed for start capacitors. While it is quite possible to use the capacitor types described above, commercial motor start capacitors are usually a special type of electrolytic capacitor which are suitable for intermittent use on AC supplies. They are smaller and less expensive than similar capacity run capacitors.

These electrolytic capacitors consist of two strips of specially processed aluminium foil interleaved with a porous separator which is impregnated with a conducting electrolyte. This is rolled up into the familiar 'Swiss roll' construction and housed in an aluminium or plastic tube. Examples are shown in Figure 8.4.

In manufacture the aluminium foil is coated with an extremely thin anodic film of aluminium oxide. This film can withstand several hundred volts across it providing the aluminium is always positive and the surrounding conducting electrolyte is negative. If the polarity is reversed the film behaves as a low resistance when only a few volts are applied to it. Because the film is so thin the area needed for a given capacitance is very much smaller than the amount needed for the equivalent paper or film capacitor.

If only one of the strips is coated in this way it can be used as a capacitor provided there is enough DC voltage present to ensure that the electrode polarity never reverses. Almost all the electrolytic capacitors used in electronic and general-purpose applications are of this polarised type and cannot be used directly in motor applications.

The reversible type used for motor start capacitors has both strips coated in

Fig. 8.4 *Electrolytic capacitors*

this way so that one strip acts as the capacitor on the positive half cycles of the AC supply and the other one takes over on the negative half cycles. The internal AC losses in this type of capacitor are much higher than in a paper or film capacitor. The capacitor will overheat if operated at full ratings for more than a few minutes but this is more than sufficient for motor start purposes.

These are special capacitors, not much used for other purposes and not easily available as over the counter sales. Mail-order suppliers are listed in Appendix 5.

The alternative, generally available polarised electrolytic capacitors cannot be used singly for motor starting but it is quite possible to use them series connected as back-to-back pairs. The arrangement is shown in Figure 8.5. The negative electrode of each capacitor replaces the two electrodes of the non-polar capacitor and the linked positive terminals connect the two electrolytes together. The rectifier diodes across each capacitor steer the positive and negative half cycles to the correct capacitor and prevent more than a fraction of a volt reverse polarity appearing across either capacitor. As far as the motor is concerned the two capacitors are electrically in series so the effective value is half the individual value i.e. two 200 μF 350V DC capacitors would be required to replace a single 100 μF 250V AC non-polar capacitor.

In operation this results in the two positive terminals charging up to the peak value of the applied voltage. This is a nuisance because the capacitors may retain their charge for several hours after the supply is removed. The high-value discharge resistors connected across

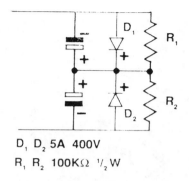

D₁ D₂ 5A 400V

R₁ R₂ 100KΩ ½ W

Fig. 8.5 *AC operation of electrolytic capacitors*

each capacitor do not affect normal circuit operation, but reduce the time taken for the capacitor to discharge to a safe voltage.

The rectifier diodes are not essential items and are only included to ensure that the published ratings of the capacitor are never exceeded i.e. never operating with polarity reversed. With the diodes present and the capacitors rated for at least 350V DC operation they will be operating wholly within their normal continuous voltage ratings and for short periods at least, could even be used as emergency run capacitors.

8.3 Metal-oxide varistors
Semiconductor devices used in motor control systems can easily be permanently damaged by transient over-voltage even if this over-voltage only exists for less than a thousandth of a second. Metal-oxide varistors are devices that are placed at key points in the circuit to prevent excessive voltages reaching the sensitive semiconductors.

They are made in the form of small round plates of ceramic-like material with electrodes metallised on each side of the plate. The construction and appearance is almost exactly the same as some types of ceramic capacitor. At the rated operating voltage the device behaves as a very high resistance and passes very little current but as soon as this voltage is exceeded, large currents flow which limit the transient voltages to a safe level.

8.4 Test meters
Most of the setting up and operation of common workshop motors can be carried out with little or no test equipment. However, there are a few items that are well worth having if you are considering something more ambitious than installing a standard motors in their normal location.

The most generally useful item is a multimeter capable of measuring voltage, current and resistance. The AC and DC voltage and current ranges should cover up to at least 500V and 10A. The resistance ranges will mostly be used for checking values of a few ohms or tens of ohms, but should also be capable of reading up to at least several megohms for checking windings to case insulation.

Small analog meters, similar to the type shown on the right-hand side of Figure 8.6, are quite low-cost items and meet most of the requirements. Their principal limitations are poor accuracy – rarely much better than 5% and only the more expensive ones include AC current ranges. However, quite a lot can be done with these little instruments and they are a useful basic minimum.

Next up in the scale is the digital multimeter. These are more accurate and can be obtained in a very wide range of capabilities. Types that include AC current ranges are suitable for motor work and are available at about twice the price of the small analog meters. The better accuracy of the voltage and current ranges is useful, but the key difference is the much improved resistance measurement facility.

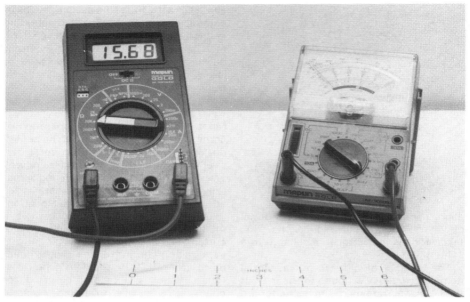

Fig. 8.6 *Digital and analog multimeters*

This is needed for checking safe motor temperatures (see Appendix 2). The resistance ranges of analog multimeters are not accurate enough for this purpose.

Additional facilities available on the more expensive digital meters include capacitance and frequency measurement. Capacitance measurement is directly useful but frequency measurement needs extra equipment before it can be used to measure motor rpm.

8.5 Speed measurement

The minimum voltage needed for frequency measurement on most digital multimeters is about 1V rms. It is possible to convert almost any small permanent magnet DC motor into an AC generator to provide this voltage but it needs a fair amount of work. The best way of doing this is to reverse the motor's normal function and rotate

the field magnets around a stationary armature. The AC output can then be taken directly from the armature coils. This eliminates the need to use slip rings to bring out the armature connections but the original motor bearings cannot now be used. The field magnets have to be mounted on a new input shaft which is carried on its own pair of bearings.

Maximum output will be obtained if the connections are made to two diametrically opposite bars on the commutator. In the common case of a three-pole armature this is exactly the same as any two segments, ignoring the third. Unless you are lucky enough to be using a high-voltage motor, more than 1000 rpm will be needed to reach 1V output. In this case remove the existing windings and rewind with as many turns as possible using much thinner wire. A good choice is 42 swg/0.1mm enamelled wire. Slightly more output is possible with 44 swg/0.08mm but 42 swg is easier to handle (Maplin BL61R and BL62S).

In the case of a three-pole armature, put all the winding on one pole and ignore the other two. In the case of a multi-pole armature, wind in a pair of slots that are exactly or, for odd numbers of poles, nearly diametrically opposite.

With a normal two-pole field magnet the output frequency will be 1 cycle per revolution of the input shaft which is 60 rpm per Hz so the measured shaft rpm will be the frequency meter reading in Hz X 60.

The above arrangement needs no additional power supply and provides an almost exact replacement for the rather expensive mechanical tachometers. However, it is limited to fairly high speed shafts. For many purposes an optical pick-off arrangement is a more convenient alternative because the measurement can be made without any direct mechanical connection to the motor under test and it is also possible to use it at lower speeds.

The system is shown in Figure 8.7. Q1 is a photo reflective infra-red sensor. This consists of a photo transistor mounted directly alongside an infra-red light-emitting diode (LED). The photo transistor is normally non-conducting but if a reflective surface is placed near the working face the reflected emission from the infra-red LED turns the photo transistor 'on' which drives the output negative.

If, as the shaft rotates, the sensor sees alternate black and white surfaces it will deliver square pulses which can be connected directly to the digital multimeter. The best contrast is between the diffuse reflection from matt black and matt white painted surfaces. Nearly as good, and much more convenient is the contrast

between black and white PVC sticky tape — strips can be applied directly to the motor shaft. Equal black and white sectors is the ideal distribution but it is not at all critical — even a 5:1 ratio is acceptable. The sensor is specified to operate at a working distance of 0.05"/ 1.3 mm but with good black/white contrast larger clearance distances are possible. Operation is easily checked by measuring the change in DC voltage at the test point as the sensor view moves from a black to a white surface.

One point that should be watched is that substances that are black and white in visible light are not necessarily black and white at infra-red wavelengths. Most whites seem to be OK but blacks can be a bit variable. A true black that has a shiny surface can also reflect a significant amount of light. Because this is a specular (i.e. mirror-like) reflection this generates additional false 'whites' which result in variable false high rpm readings which are sensitive to small changes in the working distance. With experience this can be avoided by a small change in the sensor position but it is usually better to cover the shiny surface with a dab of quick-drying photographic matt black retouching paint.

The equivalent of the AC generator set up can be achieved by using the photo sensor to monitor the speed of a separate mounted disc or drum painted with black and white stripes. If 60 stripes are used the meter will now read rpm directly and it will also be possible to measure much lower shaft speeds.

An alternative method is to use a photo interrupter-type sensor — RS Components stock No. 306—061 is a suitable type. In this sensor the LED is mounted directly opposite the photo transistor which is now always 'on' unless the IR beam is interrupted by a slotted disc. This com-

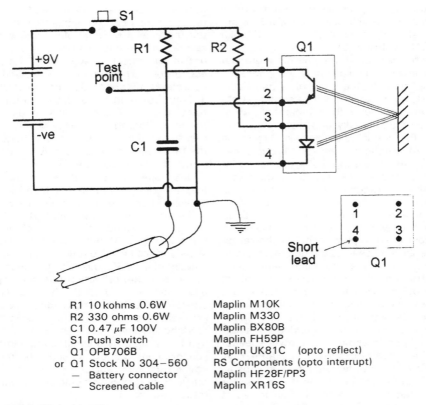

R1 10 kohms 0.6W   Maplin M10K
R2 330 ohms 0.6W   Maplin M330
C1 0.47 μF 100V   Maplin BX80B
S1 Push switch   Maplin FH59P
Q1 OPB706B   Maplin UK81C   (opto reflect)
or Q1 Stock No 304–560 RS Components (opto interrupt)
   – Battery connector Maplin HF28F/PP3
   – Screened cable   Maplin XR16S

Fig. 8.7 *Optical pick-off*

pletely avoids any specular reflection problems but it is now essential to use a slotted disc or drum beam interrupter and it is no longer possible to avoid a direct mechanical connection to the motor.

If you are checking a motor that is operating from a speed controller, switching frequency pulses radiated from the speed controller may interfere with the digital frequency measurement. To avoid this, screened cable should be used for the lead from the sensor to the digital

meter and the outer metal screen of this cable connected to the earthed metalwork of the motor and controller.

With the above methods the speed measurement can be directly read from the digital multimeter frequency readout and no special calibration is needed. An alternative method is to simply measure the DC output voltage of a small permanent-magnet motor. If driven from the shaft under test, the motor acts as a tacho–generator and the voltage output is linearly proportional to speed. The output voltage per thousand rpm will depend on the motor type chosen and the setup will need to be calibrated on a shaft turning at one or more known speeds. The most readily available known rpm is the no-load speed of almost any single- or three-phase induction motor. On full load, the rpm may be 4 to 10% below synchronous speed but at no-load the speed rises to about 1% below synchronous speed and this can be used as a reliable calibration speed.

A shaft speed can be directly checked by placing a white marker on it and illuminating it with either a 240V neon lamp (i.e. a neon lamp with built-in resistor for 240V operation) or an LED operated from half wave rectified AC. The neon lamp emits light pulses at 100 per second (= 6000 rpm), the LED rate is 50 per second (= 3000 rpm). The single white marker will appear stationary when the shaft is rotating at the light pulse frequency or a higher exact multiple. If the shaft rotates at a lower exact sub multiple the light will flash more than once per shaft rotation and multiple white markers will be visible — two at half speed, three at one third speed etc.

CHAPTER 9

Some Typical Workshop Applications

9.1 Background

Over a rather large number of years my workshop has accumulated a very useful selection of power tools and machines. Some of these are of quite venerable age but still fully functioning and rendering excellent service. One of the good things about older machines is that they were invariably built with more than adequate safety factors on section strengths and sizes and this makes them easier to repair and overhaul than their modern equivalent.

Many of these items were originally broken or scrapped and the repair and refurbishment included motor conversion or the fitting of replacement motors. Since other home workshop enthusiasts may have similar requirements the following examples of this work are included to give guidance on suitable motors and conversions for different types of machines.

9.2 10" Swing Southbend lathe

This massive 1942 vintage machine was rescued from a factory update program before it reached the scrapyard. It was originally fitted with a ¾ hp three-phase motor. This was initially replaced by a large 24V commutator motor which once

drove the variable delivery hydraulic pump in a Boulton Paul gun turret. Power for the motor was obtained from a 24V 40 amp battery charger and taps on the charger gave a limited range of speed control. The 40 amp charger occupied valuable workshop space but earned its keep because it also supplied a pre-war vintage pillar drill and a small grinder, both of which were fitted with ex-aircraft 24V commutator motors.

The low-voltage motor set up gave many years of useful service but was eventually retired when I came across a 110V DC compound-wound lift motor and a large Variac (see Ch. 5, section 5.3.5) both ripe for recycling. With the addition of a bridge rectifier this made an ideal variable DC source to drive the motor armature. The 110V shunt field was supplied directly from 240V AC via the voltage halving rectifier circuit shown in Figure 9.1.

This slightly unusual rectifier circuit delivers an average of $0.45V \times 240V$ to the field coils. A full explanation of how this rectifier works is beyond the scope of this book, but as a rough guide D1 supplies power to the field on the positive half cycles of the supply while D2 absorbs the negative inductive overswing and

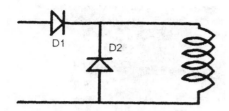

Fig. 9.1 *Voltage halving field circuit*

allows the current to continue during the negative half cycle of the supply. This circuit only works properly if it is driving a load that is an almost pure inductance so that field coils are an ideal load. The voltage halver will work after a fashion if it is used to drive a motor armature but an armature is not a pure inductance — the voltage regulation will be dreadful, and on light loads the output voltage may more than double.

Although the motor is compound wound it still takes a very high peak starting current during the small fraction of a second needed for the shunt-field current to build up to its full value. To avoid this problem a small 12V relay is included with its coil in series with the field circuit. The contacts on this relay do not close until the shunt-field current has built up to working level. These contacts are wired in series with the main armature contactor coil so that this cannot apply power to the armature until the shunt field has reached working level.

The Variac plus rectifier system is good for 10:1 speed variation at light loads dropping to about 3:1 range for heavy cuts. I still use back gear occasionally, but the six-speed belt shift is left permanently in its lowest speed setting. At maximum speed the lift motor is running at about 7,000 rpm at more than twice its normal armature voltage which must be close to its maximum permissible speed and voltage. However it has survived this treatment for more than ten years so I am well satisfied with the set up.

If you are tempted to experiment along these lines with largish motors be very sure that the motor is properly secured and protected. There is a lot of energy stored in a heavy armature spinning at 7,000 rpm and if the windings break loose much of this energy is transferred to the motor frame. The motor fixings and protection must be strong enough to prevent the motor or any broken bits from flying round the workshop.

9.3 Schaublin 13 toolroom mill
When acquired this was fitted with three, six-terminal three-phase motors — a 2 hp main spindle drive motor plus 0.3 hp and 0.1 hp motors for the table traverse mechanism and the suds pump. The motors were reconnected in delta for 240V operation and a box containing three separate single phase to three phase converters mounted out of harm's way at the back of the machine.

It would have been possible to use only one converter with component values chosen for 2½ hp operation. This is sufficient power to operate the spindle drive motor plus both the suds pump and traverse motors provided the spindle drive motor is always in circuit. It would not be possible to run the pump or traverse motors without also running the spindle motor as the start and run capacitor values in the phase converter would be quite unsuitable for a load of less than ½ hp. This was felt to be an unreasonable restriction so a separate converter was used for each of the three motors.

Most of the bits for the converters came out of my electronics junk box. The run capacitors were rectangular paper

107

dielectric capacitors, ex-World War 2 radar transmitters. The start capacitors were back-to-back pairs of ex-TV electrolytic smoothing capacitors.

The Schaublin 13 is usually operated with a very solidly built vertical milling head which is fine for general purpose milling operations but has no quill feed for sensitive drilling and jig boring operations. An alternate light-duty quill-feed vertical head is available but it is an expensive item and changing over and realigning the alternate heads is a fairly lengthy operation.

To solve this problem I use a permanently attached offset quill-feed head. The mill is normally operated with the heavy-duty vertical head fitted and aligned to be truly vertical. A separate homebrew, quill-feed drilling spindle is bolted to the side of the main milling head and displaced 3.250" to the right (Fig. 9.2). This makes both facilities immediately available interchangeably without loss of the datum setting on the workpiece.

To keep the size and weight of this attachment to a minimum, a high-speed permanent magnet commutator motor is used as the power source. This powers the spindle via a two-stage, three-speed timing belt drive. The commutator motor is much lighter than a 50Hz induction motor of similar power rating and the timing belt drive can operate with short centre distances which helps to make the head very compact. This makes it possible to carry the whole motor/speed reducer assembly directly on the quill-feed column with the weight counterbalanced by a 4-ply constant force Tensator spring.

The motor is an ex-aircraft 28V DC machine and a limited range of speed control is obtained by operating it from a transformer and rectifier unit. The transformer has 10V, 12V and 27V secondaries

Fig. 9.2 *Schaublin 13 drill head*

and these are selected by a 2-pole, 4-way switch to total 22V, 27V, 37V and 49V. These are not carefully selected voltage intervals but just a useful way of utilising an existing transformer. No attempt was made to use voltages below 22V because this was the lowest voltage at which the motor could supply sufficient torque to drive the larger drill sizes.

9.4 Jones & Shipman 540 surface grinder
Surface grinders are not common items in small workshops but they are very versatile machines and with a few extra attachments can carry out a wide range of work. Mine is used fairly frequently for surface preparation and finishing but it also doubles as a centre grinder, a slideway grinder and as a tool and cutter grinder.

The main spindle is driven by a three-phase 1 hp motor with a smaller ½ hp motor to power the hydraulics. It is an old machine and both motors are of the 415V three-terminal variety so reconnection for 240V would have entailed removal of both motors from the machine and surgical operations on the windings (see Appendix 1). This was my first essay into phase conversion and with no previous experience of modifying motors I decided to keep the machine in its standard 415V configuration and operate it from a home-brew 240V single-phase to 415V three-phase converter.

The most difficult item in this type of converter is the input auto-transformer needed to convert 240V single phase to the 415V single phase needed to drive the motor and the phase shifting capacitors. Fortunately this was in the early 1970s when conservatively rated old welding transformers could still be found, wound with copper wire and assembled with nuts and bolts securing the laminations. It was fairly easy to strip one of these down, replace the 50V welding secondary with a 175V additive overwind and re-assemble. This is not now an option because, in modern welding transformers, the laminated core joints are welded in position and it is no longer possible to strip them down and rebuild.

The capacitors were not a problem as the values needed for 415V star operation are only one-third of the value needed for the 240V delta connection. Commercial 440V power correction capacitors were used. The 1 hp and ½ hp motors, used singly or together, presented a load range of ½ hp to 1½ hp which was too wide a range to permit a single pair of start and run capacitors. This was solved by including an unloaded pilot motor which was first run up to speed using its own start and run capacitors. This then provided a balanced 415V three-phase output with sufficient power to start either or both of the surface grinder motors. Each of the surface grinder motors was provided with its own run capacitor wired straight across the motor terminals so that in any of the possible run configurations the optimum value of the run capacitor was always in circuit.

The pilot motor is a 3 hp, 2850 rpm machine and the only load on its shaft is a small fan which keeps the modified welding transformer cool. A much smaller motor could have been used but this happened to be the size I had. The large pilot motor provides lots of starting power and also improves the system balance by reducing the minimum to maximum load ratio.

With hindsight it would have been much simpler to delta connect the motors and use separate phase converters as I later did with the Schaublin 13. However, I had most of the bits for the 415V converter on hand and it had the attraction that it could be used at a later date to power other three-phase machines.

The main grinding head fitted to the J & S 540 is an excellent tool for the job that it was designed for but it is not suitable for wheels less than 5″ diameter (the speed is too low and the drive mechanics are too large) and it cannot be tilted. For some jobs, particularly slideway grinding and some types of cutter grinding, a tilting high-speed head is a great advantage. This has to be small enough and light enough to bolt to the existing cast iron wheel guard and it also needs to be vibration free and develop enough power for reasonably speedy stock removal.

The grinding head made for this job is shown in Figure 9.3. Once again it is based on surplus aircraft equipment but this time the motor used is a ½ hp,

11,000 rpm three-phase high-frequency induction motor. These motors, first used in early post-war military aircraft, can develop extraordinary amounts of power in a small frame size. The snag of course is finding the necessary 115V 400Hz three-phase power needed to drive them. This was solved by yet another surplus item — a type 8a 28V DC to 400Hz three-phase rotary converter and powered in this case by the original 24V 40 amp battery charger. The type 8a is only rated to deliver 350W which is not enough to fully power the ½ hp load but fortunately it is a rugged military machine and fairly tolerant to reasonable overload. It has led a hard life for some years now and nothing has burnt up yet!

The most recent upgrading of the J & S 540 is on the vertical feed system. The standard manual feed mechanism will easily and repeatedly advance the cut in 0.0001" increments, but to achieve this resolution the feed rate is only 0.020"

per revolution of the feed dial. The total vertical travel is 11" and this distance requires 550 turns of the feed handle which is quite an arm-aching process !

A rapid traverse motor drive is the obvious solution but it is essential that the drive does not degrade the sensitivity and accuracy of the manual feed. It is also necessary that the motor drive system is tightly controlled and does not suffer from overshoots that could result in the grinding wheel crashing into the workpiece.

The normal solution to this problem is to use a constant-speed motor with a mechanically or electrically operated clutch to engage and disengage the drive. This works fine but it entails a lot of work to manufacture the necessary motor drive and clutch arrangement. To avoid this, a relatively large low-speed high-torque motor was used which made it possible for it to be permanently connected it to the manual feed dial by a

Fig. 9.3 *Surface grinder high-speed grinding head*

low-ratio direct belt drive. At this low reduction ratio, when power is not applied to the motor, the armature rotates freely and places little additional load on the manual feed.

The motor used is an ex-washing machine commutator machine. This doesn't sound promising as it is normally classed as a high speed, fairly high power machine — about 8,000 rpm and ⅝ hp — but this is another example of the way in which the characteristics of a commutator machine can be matched to a particular application by suitable choice of the armature and field operating conditions. We need the torque that this machine can deliver but at a lower approximately constant speed — about 1,200 rpm.

In its original washing machine application, the motor is operated as a series-wound 240V AC machine. In this new application, to obtain the low constant speed characteristic, it is operated as a low-voltage (40V) shunt-wound DC machine. A small transformer rectifier unit provides a 40V DC armature supply and a separate 6V DC output supplies the field current. It would, of course, have been possible to eliminate this low-voltage field supply by rewinding the field coils with thinner wire to operate directly from the 40V armature supply. However, it was much less work to generate a separate 6V field supply by piling additional turns on top of the windings of an existing 40V transformer.

To provide the belt drive with minimum modification to the basic surface grinder, an extremely narrow (⅛" wide) size XL timing belt drive is used directly between the motor shaft and a 6¼" pulley mounted on the periphery of the manual feed dial — Figure 9.4. A 14-tooth pulley is used on the motor but, because the ratio is well over 3:1, teeth are not necessary on

Fig. 9.4 *Surface grinder fast-feed motor drive*

the large pulley and the belt runs in a simple flat-bottomed groove. The ⅛" wide belt is made by slitting a standard ⅜" wide belt with a sharp knife and snipping the embedded helically wound steel cable with a pair of cutters.

The one remaining problem is that, with any traverse distance greater than about ½", the feed dial overruns by about two full turns after the motor power has been switched off. A simple way to fix this is to keep the field permanently energised and to use a relay to short circuit the armature in the 'off' switch position. This electric braking almost instantaneously stops the feed but has dire effects on manual feed operations. The motor is now acting as a short-circuited generator and strongly resists any rapid movement. The manual feed feels as if it has been dipped into a pot of

111

extremely stiff glue!

This could be cured by an additional switch to remove the field when using manual feed but a more convenient arrangement is shown in Figure 9.5. The essential addition is C1 connected across the field circuit.

When the traverse switch is in the central 'off' position the armature is short circuited and no power is applied to the field. In the absence of any field current there is no electric braking and normal manual feed is possible. When the traverse switch is in the 'upfeed' or 'downfeed' position, the armature short circuit is removed and power applied to both armature and field. However, when the traverse switch is returned the 'off' position, although external field power is removed, the charge remaining on C1 keeps field current flowing for the small fraction of a second needed for the combined electric and friction braking to bring the motor to a halt with less than 0.001" overshoot. The field current decays to zero in less than a second which then allows unimpeded manual feed.

9.5 Paint spray compressor
This is one of the most demanding applications for converted three-phase motors.

It is not recommended unless you are prepared to make the modifications necessary to reduce the large peak starting torque that is needed when the compressor restarts while feeding an already pressurised air receiver.

When a compressor is operating and delivering air to a spray gun, the pressure in the cylinder only reaches the system operating pressure over the latter part of its compression stroke. The flywheel effect of the rotating motor rotor averages the peak torque corresponding to this system pressure over the whole intake and compression cycle, so that the effective average torque load on the motor is less than one-quarter of the peak demand of the compressor.

When the motor has been stopped, at restart the averaging effect is no longer present and in the first revolution the motor has to supply the full-peak compressor torque demand. In addition to this, while the compressor is stationary, some high-pressure air from the air receiver will leak back through the rarely perfect compressor delivery valve so that the initial pressure in the cylinder will rise well above atmospheric pressure and may approach system pressure. This causes the peak torque demand to occur

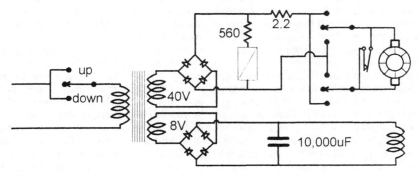

Fig. 9.5 *Surface grinder fast-feed control gear*

112

earlier in the compression cycle and increases the duration of the peak.

To be sure of trouble-free restarting at full system pressure the motor needs a starting torque at least four times normal full-load torque and, unless an excessively large motor is used, this is too much for the simple three-phase to single-phase conversions described in this book. This performance is possible if the necessary increased balanced starting currents are supplied by a large pilot motor as described in section 9.4, but it is much easier to modify the compressor system to reduce the demands on the motor.

The most effective scheme, often used commercially, is a compressor unloading system that uses a solenoid which only allows the inlet valve to close when air delivery is required. The motor runs continuously but, because the compressor inlet valve is always open, the pressure in the cylinder never rises above atmospheric and there is no significant load on the motor. When air delivery is required the inlet valve is allowed to close in the normal manner. This immediately restores the normal compressor peak torque demands, but since the motor is already running at normal speed this is not a problem.

Not all compressors lend themselves to this solution — the inlet valve may be an inaccessible reed type or may be just a row of ports in the cylinder wall near the bottom of the stroke. However, with some additional items to reduce the initial compressor torque peaks it is still possible to drive these with a phase converted three-phase motor.

The drive motor in my home-brew rig is fairly generously sized at 1 hp with a 40 μF of run capacitor provided by two large 20 μF rectangular capacitors adjacent to the motor. The start capacitor is a 100 μF reversible electrolytic capacitor

rescued from a burnt-out single-phase capacitor start motor. During the starting period this is switched in parallel with the run capacitor by a motor current operated relay so that the total value of the start capacitance is 140 μF. This current-controlled system is one of the alternative methods of starting three-phase motors described in *Electric Motors* (Nexus Special Interests). However, the method described in Chapter 2 of this book is equally suitable and has the advantage that it does not need a specially modified motor current relay.

When the compressor is operating, the cooled output from the compressor delivery valve contains condensed water vapour and a little oil, both in the form of a fine mist. A condensate filter in the line removes this from the high-pressure air before it reaches the main air storage receiver. In commercial compressors this filter is a fairly small item and is provided with a manually operated valve at the lowest point to drain away accumulated tramp oil and water condensate.

In the home-brew system, as part of the reduced starting torque strategy, the volume of this item is made quite large — more than ten times the compressor displacement. It is connected to the main air receiver via a non-return valve. An automatic pressure dump valve releases the pressure in this filter volume whenever the motor is switched off.

When the motor restarts, the compressor now delivers air into the empty filter volume. Long before the pressure has built up in the filter to full system pressure the motor has reached its full running speed and can handle the full average torque load without difficulty.

The additional items required are the oversize condensate filter, the non-return valve and the pressure dump valve.

The condensate filter is simply a length

of 3" pipe provided with "O" ring sealed end caps retained by a central stud running the full length of the filter. The filter is mounted with its long axis vertical and the top two-thirds tightly packed with knitted stainless steel pot scrubbers from the local supermarket. Air enters at the bottom and exits at the top. The pressure dump valve connects to the bottom of the filter so that every time it operates, it vents any accumulated oil and condensate.

The non-return valve is a simple manufactured item using a ½" diameter stainless steel ball, spring loaded into contact with an "O" ring seating face. The reverse leakage of this type of valve is almost nil. When the compressor is used as an ever-ready workshop-compressed air supply the air receiver retains its pressure all day without topping up.

The pressure dump valve is a completely standard 240V AC solenoid-controlled water inlet valve removed from a junk automatic washing machine. Although designed to control water flow it seems to work equally well on compressed air even when (or perhaps because!) this is contaminated with condensate and tramp oil. The solenoid cannot be connected directly across the motor terminals because the dump valve opens when the solenoid is energised. Instead, an additional relay is used which is energised when power is applied to the motor. When it is energised the normally closed contacts on this relay open and remove the power from the solenoid. This closes the dump valve and allows pressure to build up in the system.

9.6 10" Twin-wheel grinder

This is a really ancient machine rescued from the breakers yard (Figure 9.6). The main shaft runs on two substantial white metal bearings in a heavy solid bench mounting casting. It came fitted with a pair of flat belt fast and loose pulleys for drive from overhead shafting but these were replaced by a single vee belt pulley.

I initially powered it with a ½ hp, 1,425 rpm split-phase motor with a step-up vee belt drive to run the twin 10" by 1" wide wheels at 2,300 rpm. This is not the right type of motor for this sort of duty — it takes several seconds for the motor to run the heavy twin wheels up to operating speed and this is far too long for a split-phase motor. A single start is OK but two or three successive starts will seriously overheat the start winding unless ample time is allowed for the windings to cool down between starts. A capacitor-start motor is the right type of single-phase motor for this sort of duty, but since I did not have one I used the split-phase machine and tried to treat it gently.

After a couple of years the inevitable happened and the starting winding burnt out. However, by this time I had worked out the techniques for converting three-phase machines and this was the obvious solution. The local scrapyard yielded a 1 hp six-terminal machine which fitted nicely because it was about the same size as the previous ½ hp machine and there was plenty of room for the usual phase conversion components which fitted in a small box beside the motor.

The starting torque of the converted three-phase machine is rather better — the wheels now reach operating speed in less than two seconds — but the important point is that the whole of the three-phase winding is used during the starting cycle. This takes much longer to heat up than the relatively small amount of copper (about $\frac{1}{10}$) which makes up the start winding of a split-phase machine, so the three-phase motor is very much more tolerant of long starting periods.

Fig. 9.6 *10" grinder*

9.7 Power fretsaw

This machine doesn't need a lot of power to drive it but variable speed is a useful asset to enable it to be used for both woodworking and metalworking. The small commutator motors used to drive domestic spin driers are ideal for this sort of job.

Figure 9.7 shows the set up with one of these motors driving a small fretsaw. A 12:40 ratio, size XL timing belt drive is used and this is a good example of the short centre-to-centre distance that is possible with this type of drive. A collector shoe picks up a little air from the centrifugal cooling fan on the motor and this is used to blow swarf clear of the cutting edge.

The motor is operated from the simple triac controller described in Chapter 5 and has more power than is really necessary. Because of the very light load on the motor this gives a useful speed range of about 4:1 and most operation is in the lower part of the coverage.

9.8 Linisher

Figure 9.8 shows a small linisher powered by an ex-dishwasher main pump motor. Once again a size XL timing belt is used for the drive. The motor has ample power for this application and the 2,850 rpm shaft speed is convenient as it allows a near 1:1 ratio drive for the abrasive belt speed of about 1,800 ft/minute. The dishwasher motor is a capacitor-run type

Fig. 9.7 *Power fretsaw*

$(7.5\,\mu F)$ which has poor starting torque but, as the linisher always starts with no applied load, this is not a problem.

9.9 Pillar drill
This a Startrite Mercury ½″ pillar drill fitted with a ⅓ hp single-phase motor

Fig. 9.8 *Linisher*

driving the quill through a four-step pulley vee belt drive. It is a good general-purpose machine, but the lowest speed setting is not low enough for large diameter fly cutting and shifting the belt on the four-step pulleys for the normal speed changes means messing about with a spanner and retensioning the belt.

Eventually I got fed up with this and fitted it with a variable speed commutator motor drive. The set up is shown in Figure 9.9. One of the larger ex-automatic washing machine motors is fitted with a 32-tooth XL flanged timing belt pulley and this drives an 8" diameter Tufnol pulley bolted to the top of the original four-step, motor-mounted vee belt pulley. Out of sheer laziness I retained the original ⅓ hp drive motor although its only remaining function is to provide a pair of bearings for the four-step pulley. The drive ratio is just over 1:4 so no teeth are needed on the large pulley. Tufnol was used simply because my scrap box happened to have a disc of the right size. A pulley made from varnished ½" thick plywood would have been quite OK. Either way, it's a lot quicker and easier to make than the equivalent 125-tooth pulley.

After much head scratching, I managed to sort out the connections to the particular speed control board fitted to this washing machine and this is housed in a small control box mounted on the left of the drill head. With four vee belt drive ratios and a wide-range, fully variable motor speed control I expected it to be easy to lose track of spindle speeds so I also fitted a simple tachometer readout

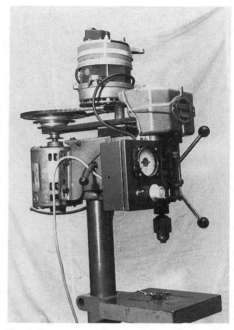

Fig. 9.9 *Pillar drill drive (belt guard removed to show drive arrangement)*

in this box.

In fact, the variable speed control is so effective that I only ever use the two lower speed vee belt settings and, apart from fly cutting, it stays permanently on the second lowest setting. The variable speed facility is a great convenience. The only real drawback is the noise level — at maximum speed a largish commutator motor screaming round at 8,000 rpm at close range is far from peaceful!

APPENDIX 1

Converting Three Terminal Three Phase Motors

This procedure is only included as an appendix because not all three terminal motors are suitable for conversion. Some may have the windings so heavily dunked in resin that the junctions are inaccessible and some of the larger sizes may already be delta connected. However, about 80% of these older motor types can be successfully converted so the attempt is well worthwhile.

Remove both end bells from the motor. This will expose the three leadout wires bound in to the end turns of the windings with a mess of string, tape and resin. The star point is usually buried beneath the joints between the leadout wires and the windings. Carefully cut through all the string and tape and gingerly prise one leadout wire free of the windings. The junction between the leadout wire and the wire of the motor windings will usually be covered with a short length of black or yellow sleeving.

Check that only one wire emerges from the other end of this sleeving. This means that the windings are star connected. If you're really unlucky you may find two wires emerging — this means that the windings are already delta connected, there is no star point, and the motor is only suitable for 415V operation. If this

is the case, your only choice is to either scrap the motor or to buy a commercial phase converter. Don't be too discouraged by this possibility because delta-connected windings are quite rare in the power range likely to be used in the home workshop.

Once you have identified the junctions to the three leadout wires look for the fourth piece of sleeving — this will be the star point where the three inner wires of each winding are joined together. This is the first joint to be made in the assembly of the motor, so it is usually buried under one of the other three junctions. Very carefully cut or scrape away the resin impregnated sleeving until the twisted and soldered joint between the three wires is visible. Don't attempt to unsolder the joint — this is too fiddly a job — but cut each wire just before the soldered joint. Scrape each of the three ends to bright clean copper and twist and solder to three new leadout wires. To simplify sorting out the final connections, use a different colour for the new leadout wires so that they cannot be confused with the original wires from the outer ends of the windings.

Occasionally each of the main windings is not a number of complete turns but includes an extra half turn. In this case

119

the star point is not buried under the leadout junctions but appears as a single, three-wire joint at the other end of the stator. In this case the new leadout wires can be attached directly to these three wires and it is not necessary to disturb the original leadouts at the other end of the stator.

The maximum temperature of the windings is too high to risk using friction tape or any of the ordinary PVC or cellulose sticky tapes. High-temperature electrical grade Mylar or glass fibre tape is OK but only available from specialist suppliers. The simplest solution is to bind with half a dozen layers of the very thin PTFE/Teflon tape used by plumbers instead of pipe jointing compound. This is available at any central heating stockist. Once each of the three new joints is safely insulated continue the tape wrapping up the first inch or so of the new leadout wires so that the wires will still be protected even if the temperature gets high enough to soften the PVC.

Tie the six leadout wires to the end turns of the motor windings with string and secure the string and any potentially loose ends or wires with two-part epoxy resin. Pair each inner wire with its respective outer by checking continuity with an ohmmeter. Twist each pair together and connect to the terminal block as shown in Figure A1.1. Because the three windings are identical, the order in which the windings are connected to the block doesn't matter. The only point to watch is to be sure that each terminal carries the inner of one winding and the outer of another.

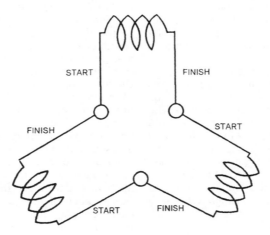

Fig. A1.1 *Delta winding connection*

APPENDIX 2

Motor Power Rating

Section 1.5 in Chapter 1 included some comments on the nameplate rating of induction motors and permissible overloads for intermittent working. This is fine when an informative nameplate is fitted but many useful motors carry no nameplate, and power rating is a matter of guesswork. Size alone is not a reliable indicator. Accurate assessment is not possible without extensive testing and expensive test equipment, but the following notes should enable estimates to be made that are good enough for most requirements

The maximum torque that a motor can deliver is mainly controlled by the volume of the active part of the rotor/armature and not much affected by the other dimensions. A large overall diameter stator/field assembly may improve the efficiency and reduce the temperature rise of a motor but has little effect on the maximum torque that can be delivered from a particular size of rotor/armature. Low-speed motors (up to about 3,000 rpm) can utilise a roughly constant fraction of this torque, but at higher speeds unwanted losses in the iron increase and it is necessary to accept a lower torque rating to avoid excessive temperature rise. The amount of the reduction depends

greatly on motor design and cooling but only one-third of the low-speed value may be available at 24,000 rpm.

Comparisons based on rotor/armature size are useful but only if the rated speed is either known or can be measured.

An alternative approach relies on the fact that, for motors operating at shaft speeds of up a few thousand rpm, reasonable guesses can be made based on the fact that the fraction of the maximum input power that is dissipated in the winding resistance is roughly constant. Providing the normal operating voltage of the motor is known, estimates can be made based on the measured winding resistance. Figure A2.1, reproduced from *Electric Motors*, shows this in graphical form for commonly encountered supply voltages. No great accuracy can be expected from this very simple method because it assumes that all motors have roughly similar distribution of losses. Nevertheless, it is a useful first approximation and will give consistently better results than guesses based only on overall size and weight.

The method is not suitable for high-speed commutator motors because the distribution of losses is different and the indicated power would be unduly opti-

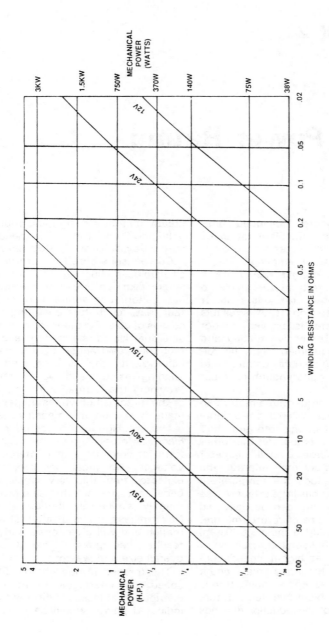

Fig. A2.1 *Motor power rating*

mistic. However, it is a reasonable basis for low-speed commutator motors and the normal range of single- and three-phase induction motors.

To avoid the variable contact resistance of the carbon brush contacts, armature resistance should be measured directly across the appropriate commutator bars. Series- and shunt-field windings should be ignored.

In single-phase motors, measure only the resistance of the main 'run' winding. With capacitor-start or capacitor-run motors this is the resistance across the motor terminals because the DC connection to the start winding is blocked by the capacitor. With split-phase motors the start winding is normally connected across the main winding via the centri-

fugal switch, so it is necessary to either disconnect this winding or hold open the centrifugal switch. For three-phase motors, measure across any pair of terminals.

For most workshop applications the actual power output capability is of limited interest — the key question is whether the motor will overheat when driving your worst case load. For most of the examples in this book, the motor described has ample power capability and overheating is not a problem. However, you may be faced with a case where the motor seems to be running very hot and you don't want to run the risk of damaging a hard to replace motor. In most cases your fears are likely to be unfounded — modern motors running near the limit of their ratings run very hot indeed. Subjec-

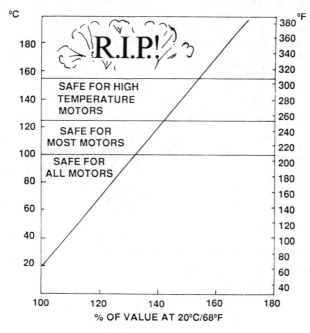

Fig. A2.2 *Change of winding resistance with temperature*

tive guessing of the running temperature, or even taping a thermometer to the outside casing, is likely to be wildly inaccurate. The only safe method is to directly measure the temperature of the windings.

Copper has a temperature co-efficient of 0.4% per degree C. If the resistance of a winding is first measured at room temperature, the increase in resistance at final working temperature is a direct indication of the actual winding temperature. Figure A2.2, again taken from *Electric Motors*, shows safe limits for the common types of motors. Use a digital multimeter to measure the resistance — few analog meters are sufficiently accurate.

This is an easy measurement on 240V single- and three-phase induction motors, but the armature resistance of commutator machines may be too low to measure accurately on even on the lowest multimeter ohms range. In this case pass a fairly large fixed DC current through the

armature and use the multimeter to measure the voltage drop between well separated commutator bars. A bar near or under each of the brushes is suitable. The set up is shown in Figure A2.3 — several amps may be needed and a starter battery or battery charger is a convenient source.

Use the 2V range of the multimeter. Set the current to give a convenient meter reading. If 1.000V is chosen, with the windings at room temperature the multimeter will directly indicate the percentage increase in winding resistance when hot. The current must be maintained at the same set level for both the cold and hot measurements. To avoid errors arising from variable contact resistance, it is essential that the multimeter is connected directly to the commutator bars and that the current flow is introduced through *separate* contact points — the main motor brushes are usually the most convenient current connection points.

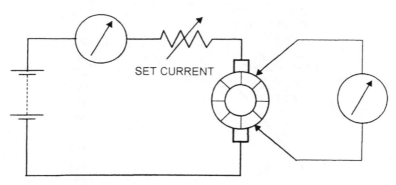
SET CURRENT

Fig. A2.3 *Low-value resistance measurement*

APPENDIX 3

Demagnetisation and Remagnetisation

A3.1 Magnet materials

Many of the older types of permanent-magnet motors used metal (Alnico, Alcomax and similar alloys) magnets. Long magnets (about twice as long as their width) were necessary because the metal magnets are not as resistant to demagnetisation as modern ferrite magnets. The highest performance could only be obtained if the magnets were magnetised after assembly with the armature in position. If the magnetic circuit was broken by removing and then replacing the armature as much as 30% of the flux would be lost and this could only be restored by remagnetising the assembled motor.

Some very high-performance servo motors still use this type of magnet because they can operate at a higher maximum flux density than ferrite magnets (i.e. produce more torque). To simplify the remagnetisation problem, these motors sometimes carry a few turns of thick wire wound round each magnet. After final assembly, a several thousand amp current pulse is passed through these windings to fully magnetise the field magnets.

Modern motors mostly use ferrite magnets — this is a black ceramic material which is much more resistant to demagnetisation than the older types of metallic magnets. Much shorter magnets can be used, length less than half the width, and it is usually possible to remove and replace the armature without significant loss of flux. This is fortunate because ferrite magnets are much more difficult to magnetise than the Alni/Alcomax family. The latter can be magnetised with a single pulse of about 8,000 ampére/turns per inch length of magnet. A ferrite magnet needs about 30,000 ampére/turns per inch.

Ferrite magnets can easily be distinguished from metal magnets by their lower density and their appearance when a small area is lightly ground. The grinding wheel will make little impression on the ferrite and the appearance of the freshly ground surface will not differ much from the rest of the magnet. The ground surface of a metal magnet will show the characteristic silvery metallic sheen.

In addition to these magnet types there are now the rare earth magnets based on samarium cobalt (SmCo) and neodymium iron boron (NdFeB) alloys. These are metallic magnets that combine the advantages of both the metal and the ferrite ceramic magnets. However, they are very expensive and will only be encountered in very high performance servo motors and similar items. They can be distinguished from the older metallic magnets by the characteristically short magnet lengths used — even shorter than similar ferrite field magnets.

A3.2 Demagnetisation problems
Armature currents exert a demagnetising force on the field magnets and manufacturers take this into account when choosing the magnet size. Large currents, limited mainly by the armature resistance, occur when the supply voltage is first applied to the motor. This current is almost doubled if the supply voltage is reversed and re-applied to a motor that is already running at full speed in the forward direction (this is because the back EMF now adds to the supply voltage instead of opposing it). Manufacturers allow for this rather brutal treatment and, at least in the small and medium sizes, the field magnets will retain their full flux density indefinitely.

However, if the motor is being operated at say twice its normal rated voltage, the current surge at startup will already be twice the normal value and may approach four times if the motor is suddenly reversed. This is likely be sufficient to partially demagnetise the field magnets and reduce the operating field strength. If operating a permanent-magnet motor above its normal voltage rating, sudden reversal should be avoided or steps taken to limit the maximum current that can flow.

It should be emphasised that loss of flux due to demagnetisation is an effect which occurs almost instantaneously when a particular excessive current level is reached. There is no second chance and the loss can only be restored by remagnetising.

Demagnetisation often appears to develop slowly as if the magnets were wearing out. This is because of the random timing of the occurrence of peak current surges. In battery-operated systems, the worst cases occur with a freshly charged battery driving a stalled motor or reversing from high speed. Large currents flow under these conditions but the actual demagnetising effect on the field magnets depends both on the size of the current peak and on the precise position of the armature at the instant of the current peak.

On most, or all, of the occasions the combined effect is below the threshold value needed to decrease the field strength. However, if it exceeds the threshold, it reduces the field strength by an amount which depends on the position of the armature at the time of the current peak. Once the field strength has been reduced it will not be further reduced (i.e. there is now a higher threshold) until it is subjected to an even higher demagnetising peak. Over a large number of starts and reversals the field strength will eventually reach the reduced level corresponding to the worst case peak current and armature position combination.

Change in field strength can easily be monitored by looking for any change in the voltage required to maintain some chosen fixed no-load speed. Convenient speeds are 1,500 rpm or 3,000 rpm as they are easily checked with a neon lamp (see Chapter 8). With a fixed-supply voltage, no load-speed *increases* in direct proportion to the loss of flux. With the speed set to a chosen value by the neon strobe monitor the voltage measured at the motor terminals will *decrease* if there is any loss of flux.

Demagnetisation can certainly occur, but it is not often encountered in low-speed motors fitted with ferrite magnets and operated within their normal ratings. High-speed motors are a little more susceptible because they usually have the brushes slightly advanced away from the 90-degree neutral position to improve the high-speed commutation and this increases the interaction between the two magnetic fields.

A3.3 Remagnetising

To remagnetise a magnet, all that is needed is to apply a powerful magnetic field to it to temporarily raise the magnetic flux density within it above the saturation value. When the applied field is removed the flux density within the magnet falls back a little and then remains at this lower value indefinitely. The change of state occurs almost instantaneously and it is only necessary to apply the field for a few thousandths of a second.

We can take advantage of this to jury rig an extremely simple home-brew magnetiser. The normal domestic 240V AC ring main is protected by a 30A fuse and each outlet plug can be fitted with a 3A, 5A or 13A fuse. A standard 13A fuse will carry 13A for long periods. If a short circuit is placed across the outlet a very large current will flow but it takes several milliseconds for fuse wire to heat up to its melting point and break the circuit — this is sufficient time and current to magnetise the size of magnets we are likely to encounter in small, permanent-magnet motors.

The set up is shown in Figure A3.1. F1 is a standard ceramic body 13A cartridge fuse. The two rectifier diodes are rather special but readily available low-cost devices. Although the continuous rating of these diodes is only 6A their peak current rating is very high — 400A for 8mS. At this sort of current there is significant voltage drop in the domestic wiring, but with typical installations this circuit will cheerfully deliver 400A peak into a ½ ohm magnetising coil — a peak power of 80kW which is more than enough to magnetise quite sizable magnets.

At this current level it is not necessary to use an iron-cored purpose-built electromagnet. A convenient starting point is 28ft/8m of 23 swg/0.6mm PVC insulated single-core wire (Maplin PA56L). This has a total resistance of about half an ohm and is thick enough to accept a single short duration 400A current pulse without melting the PVC. Sufficient field

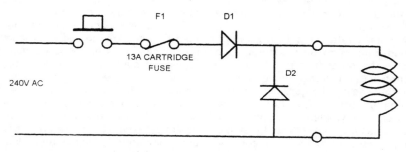

DI & D2 MOTOROLA MR754 (MAPLIN YH97F)

Fig. A3.1 *Magnetising rig*

127

Fig. A3.2 *Magnetising setup*

can be generated if the magnetising coil is wound directly on the motor or the magnet to be magnetised. Wind as much as possible on to the workpiece with the remaining wire left as a single loop completing the connection to the rectifiers. To complete the magnetic circuit, a U-shaped iron circuit should enclose one side of the coil and finish at least reasonably close to the outside casing of the motor. Any convenient lumps of scrap mild steel can be used but, for small motors, the simplest scheme is to lightly grip the coil and motor assembly in a small machine vice. Figure A3.2 shows a typical set up.

With small motors the field strength is strong enough to both saturate and penetrate the outer soft iron casing to remagnetise the field magnets contained within the casing. Larger motors may have magnets fitted within heavy cross-section cast or welded soft iron return paths. In this case locate the magnetising winding in two sections − one round each of the field magnets. An alternative technique is to remove the armature and replace it with a slug of mild steel which carries the magnetising winding.

The normal domestic 30A ring main is designed to be capable of blowing the 13A fuses fitted to standard 13A plugs − but *only* as occasional infrequent events arising from faulty user equipment. The peak current from such events will cause the ring main voltage to instantaneously drop well below normal voltage limits and this may interrupt the program operating on any computer connected to the same ring main. It is also beyond the rating of many 13A wall switches and may damage switch contacts.

For these reasons you should only resort to this technique on those very rare occasions when it is necessary to remagnetise a motor. Frequent fuse blowing activities are likely to be very unpopular with local computer operators and also, probably, the electricity supply company! Also, use only the standard ceramic cased 13A cartridge fuses that are designed to fail safely and reliably on domestic 240V supplies.

128

APPENDIX 4

Motor Terminology

Although in the main part of the book every attempt is made to explain things in everyday language it is impossible to completely avoid specialised terms and jargon. The following list of admittedly rough definitions is provided to help sort out some of the more commonly used terms.

AC Alternating Current – current that continuously varies smoothly back and forth from positive to negative values. In domestic AC supplies this occurs 50 times per second – this frequency is normally called 50c/s or 50Hz.

240V AC Used in this book to refer to the normal UK domestic AC supply. Most of Europe uses a 220V nominal supply. In the future, both areas intend to standardise on a rather wide tolerance 230V ± 10% nominal supply. This range is wide enough to cover the total normal voltage variations of both 240V nominal and 220V nominal supplies. Each area happily continues to distribute its original supply voltage range but is now entitled to call it 230V nominal whenever this is politically more acceptable!

Armature The rotating part of a com-mutator motor or dynamo.

Autotransformer A transformer in which the secondary winding is either joined to, or forms part of, the primary winding – see transformer.

Back EMF When power is applied to a motor and it rotates, it also acts as a generator and produces within it a voltage that opposes the applied voltage. This voltage is called the back EMF.

Bridge rectifier An interconnected as-sembly of four rectifier diodes. Two terminations are the AC input – the other two are the DC output.

Brush A fixed carbon or metal conductor which is spring loaded to bear on, and to make electrical contact with the commutator of a motor or dynamo; or the sliprings of a generator.

Contactor A heavy-duty relay (see relay) used for controlling high power circuits.

Closed A pair of contacts or a switch in the connected 'on' position.

Capacitor A device consisting of con-ducting surfaces separated by an insulator which can store electrical energy – see Chapter 8.

Commutator A rotating multi-way switch which controls the direction of cur-rent flow in the armature of a DC or universal motor. It usually consists of a cylindrical array of copper segments mounted on the rotating part of the motor. Each segment of the array connects to part of the armature windings and power is fed to the windings by fixed carbon or metal brushes which bear on the segments.

DC Direct Current − current which always flows in the same direction. Batteries are a source of direct current. Although, strictly speaking, the term defines a current it is commonly used to mean unidirectional e.g. DC voltages or DC currents are unidirectional voltages or currents.

Diode A semiconductor device that only allows current to pass in one direction. Higher power diodes or assemblies of more than one diode are often called rectifiers. There is no fundamental difference between a diode and a rectifier − the terms are just different names for the same device.

Displacement The volume of air (at normal air temperature and pressure) displaced by the piston moving from the bottom to the top of its stroke in the cylinder.

EMF Electro Motive Force. An almost archaic term for voltage but still commonly used when referring to the back EMF of a motor − see above.

Energised When sufficient power is applied to a relay to cause it to operate it is said to be energised.

Excited excitation When current is passed through a winding of a motor or generator, a magnetic field is generated and the winding is said to be excited.

Field The fixed magnetic field in which the armature of a commutator motor or dynamo rotates. The field can be provided by an electromagnet formed by field coils wound round an iron yoke or by permanent magnets.

Heatsink Many power semiconductor devices such as transistors or rectifiers will overheat unless bolted into thermal contact with a larger area of metal that can dissipate the unwanted heat − a heatsink.

Hz See AC.

Impedance In AC circuits the effective resistance to current flow may differ from the DC resistance because of the presence of inductors or capacitors. Impedance is the AC resistance of such a circuit.

Inductance A coil of wire possesses inductance because it can store energy in the magnetic field that it generates when a current is passed through it. On AC supplies, the alternating polarity of this magnetic field causes the apparent resistance to current flow (i.e. the 'impedance') to be much higher than the DC resistance.

LED Light-Emitting Diode. Several colours available in the visible light region but diodes emitting infra-red are more efficient.

MOSFET A type of three-terminal semiconductor that is normally 'off' but can be switched partially or fully 'on' by a voltage applied to the third (gate) terminal.

Open A pair of contacts or a switch in the 'off' position.

Open circuit Non-conducting, disconnected or without any load connected.

Parallel Parallel connection. Components (e.g. resistors, switches, motors etc.) connected across a pair of wires so that the same voltage appears across each component.

Peak Alternating current moves smoothly from a maximum value in one direction to a maximum value in the other direction. These maximum values are referred to as the peak value of the voltage or current. The peak voltage of domestic 240V supplies is 1.414 × 240 = 339V

Pot. See potentiometer.

Potentiometer A three-terminal variable resistor − the slider and *both* ends of the resistive element are available as connections.

Reed relay A relay in which the contacts take the form of two flat metal blades ('reeds') hermetically sealed into opposite ends of a small tubular glass enclosure. The reeds are made of magnetic material and this makes them stick together and switch on when current is passed through the relay coil that surrounds the glass tube.

Relay A switch or series of switches (usually called contacts) operated by an electromagnet.

Relay coil The coil of wire that surrounds the magnetic parts of a relay to form an electromagnet. When sufficient current is passed through this coil it generates a strong magnetic field which operates the contacts of the relay.

Rectifier A semiconductor device that only allows current to flow in one direction. Used to convert (i.e. rectify) alternating current (AC) to direct current (DC) which is needed by some types of motor. See also diode.

Resistor A component that impedes or reduces (i.e. resists) the flow of current round an electrical circuit. The value of a resistor is stated in ohms, kilohms (thousands of ohms) or megohms (millions of ohms).

RFI Radio Frequency Interference. Commutator motors and some electronic speed controllers may radiate radio interference, mainly in the medium and long wavebands. Check out your motor rig with a portable radio!

rms Root-mean-square. The single figure used to identify the effective value of an alternating current or voltage − in spite of the fact that the actual value is changing all the time. Roughly speaking, it is the AC equivalent of the DC value. One amp rms through a one ohm resistor would require one volt rms and dissipate a power of one watt i.e. exactly the same power as 1 volt and 1 amp DC. The peak value of domestic 240V rms AC supplies is a little more than 1.4 X the rms value.

rpm Revolutions per minute.

Rotor The rotating part of an induction motor or AC generator.

SCR Silicon-Controlled Rectifier. Sometimes called a thyristor. A semiconductor switch that works in the same way as a triac but the controlled conduction is in one direction only. In the reverse direction current flow is always blocked i.e. it acts as a rectifier. See also triac.

Series Series connection. Components (e.g. resistors, switches, motors etc.) connected together daisy-chain fashion so that the same current flows through every component.

Solenoid A special type of electromagnet in which the moving part (often a cylindrical iron slug) moves within, and is surrounded by, the coil which produces the magnetic field. Used when a relatively long movement is required.

Stalled Stationary rotor or armature. Usually when the torque generated by the motor is insufficient to turn the load.

Stator The fixed part of an induction motor or AC generator that usually carries the main power windings.

Tacho generator A device which generates a voltage or a frequency proportional to shaft speed.

Torque Turning or twisting force. May be used to describe the turning force exerted by a motor shaft or to describe the force necessary to turn the device that the motor is driving.

Transformer A device for changing the voltage of an AC supply. Consists of a primary (input) winding and one or more secondary (output) windings wound round a laminated iron core.

Transients Transient voltages and/or transient currents are extremely short duration excessive voltages or currents. They can occur as a result of switching an inductive load, such as a motor or they may appear at random times on the 240V AC mains.

Transistor A three-terminal semiconductor type that is normally 'off' but can be switched partially or fully 'on' by a current entering the third (base) terminal.

Triac A semiconductor switch which can block or allow large currents to flow into a load. When it is switched on by a small current into its gate electrode it conducts in both direction so that it is suitable for use in AC circuits.

APPENDIX 5

Suppliers

Cirkit Distribution Ltd.
Park Lane
Broxbourne
Herts EN10 7NQ
Tel: 01992 441306

Mail-order suppliers of a wide range of electronic components mainly aimed at the electronic constructor market.

J & N Factors
Pilgrim Works
Stairbridge Lane
Bolney
Sussex RH17 5PA
Tel: 01444 881 554

Retail factors – motor start and run capacitors. Battery charger parts and miscellaneous surplus motors.

Maplin Electronics
P.O. Box 3
Rayleigh
Essex SS6 8LR
Tel: 01702 554161

Retail distributors of enormous range of electronic equipment, electronic components and related supplies. UK and overseas mail order. Annual catalogue on sale at branches of the W. H. Smith stationers chain.

Power Capacitors
30 Redfern Road
Tyseley
Birmingham
B11 2BH
Tel: 021 708 2811

Start and run capacitors.

RS Components
PO Box 99
Corby
Northants NN17 9RS
Tel: 01536 201201

Major supplier of extremely wide range of electronic, electric and allied components. Mail order plus local outlets in London, Birmingham, Corby and Stockport.

Rush Industrial Sales
Biggin House
126 Station Rd
Tempsford
Sandy
Beds. SG19 2AY
Tel: 01767 640779

Helpful stockist of a very wide range of motor start and run capacitors.